the prehistoric
past
revealed

MITCHELL BEAZLEY

the prehistoric past revealed

The four billion year
history of life on Earth

Douglas Palmer

The Prehistoric Past Revealed
by DouglasPalmer

I would like to thank all the team at Mitchell Beazley for their efforts,
especially the project editor Naomi Waters and picture researcher Jenny
Faithfull who has gone to great lengths to get hold of some obscure
illustrations. I also acknowledge Robert Dinwiddie's expert help and
advice in chapter 8.

First published in Great Britain in 2003 by Mitchell Beazley,
an imprint of Octopus Publishing Group Limited,
2–4 Heron Quays, London E14 4JP.
© Octopus Publishing Group Limited 2003
Text © Douglas Palmer 2003

1 84000 736 2

A CIP catalogue record for this book is available from the British Library.

While all reasonable care has been taken during the preparation of this edition, neither the publisher,
editors, nor the authors can accept responsibility for any consequences arising from the use thereof
or from the information contained therein.

Commission Editor: Vivien Antwi
Executive Art Editor: Christine Keilty
Project Editor: Naomi Waters
Design: Miranda Harvey
Picture Research: Jenny Faithfull
Production: Gary Hayes
Copy-editor: Siobhan O'Connor
Proofreader: Lindsay Porter
Indexer: Sandra Shotter

Page 2: Fossil leaf from Messel Pit, Germany
page 5: *Caudipteryx*, small Cretaceous bird-like dinosaur

Set in DINEngschrift and Bliss

Printed and bound in China

Contents

introduction

The aim of *Prehistoric Past Revealed* is to explore the geological history of the Earth and its inhabitants from the present day with its more familiar environments and life, and work back progressively through time into the less well-known depths of the Earth's geological past.

Although this narrative goes against the flow of time and evolution, it follows the same path as the discovery of Earth history . Generally, the youngest and most recent sediments, rocks, and layers of strata are either at or close to the present land surface. These recent "pages" of the Earth story tend to be less disturbed by geological processes such as faulting, folding, or metamorphism than the older ones, simply because they have not been around for long enough to suffer such processes. At least this is the situation in northern Europe, where so much of the early exploration of Earth history took place. Elsewhere, in places such as Mediterranean Europe or California, this generalization does not hold. Young strata are highly deformed by recent Earth movements and continuing mountain building. But in both types of geological environment, fossils found in young strata are more closely related to living forms than those found in more ancient strata. The further back in time you go, the greater the proportion of the fossil biota that belongs to extinct and less familiar groups of animals and plants.

By accident of geological history, northern Europe, and especially the British Isles, contain within a very small geographical area a remarkably good sample of the last 1,000 million years or so of Earth history. Across Britain from south to north, younger to older successive strata are laid out like the pages of a book waiting to be leafed through. However, the book and its narrative was read backwards, from the most recent ie the present, towards the beginning, the most ancient. And it soon became evident that the closer to the beginning you went, the more pages and sometimes whole chapters were missing. Furthermore, the earlier pages are often torn and crumpled so that it is difficult to decipher the text.

Then it was realized that there was not just one copy of this book, but that each continental region had its own edition, a bit like the different gospels of the New Testament. In order to get the whole story it was necessary to try to match the different versions. This task is still not complete, but is being worked upon all the time.

When the geological vastness of the great continents of the Americas, Africa, Asia, Australia etc gradually became better known, it was also realized that the European version of the story was very much a minituarized version. Many of the most important geological processes and products, from river systems to mountain ranges, are normally on a scale which are an order of magnitude larger than those represented by so many European rocks. The Thames or the Rhine are small compared with the Amazon, Indus, or Mississippi. Likewise the ancient Paleozoic "mountains" of Scotland pale by comparison with the Himalayas and Andes or Western Cordillera of the Americas.

Perhaps the greatest shift in understanding of the geological past came with new investigative techniques that were driven forwards by the necessities of the Industrial Revolution. For the first time geology became a modern profession and academic subject of investigation with the establishment of national surveys and university departments. But it was the dire necessities of World War II, which drove the new techniques forward and were to bring a revolution in geological understanding. The accurate dating of igneous rocks using radioisotopes took a long time to become a reliable everyday technique. But by the 1950s, radiometric dating promoted a great leap forward, as did the use of earthquake waves to explore the inaccessible depths of the Earth, revealing its layered structure. The measurement of rock magnetism, combined with the detailed mapping of the ocean floor, all provided essential ingredients which fuelled the plate tectonic revolution.

These developments caused a radical reappraisal of the old methodologies and paradigms about the past, especially concerning the early history of the Earth. It was realized that there are many geological processes which cannot be easily assessed by what happens within a daily, annual, or even decadal scale. Many important events are very rare and others happen so slowly that we are only now recognizing their significance. Since astronomy put the Earth and its formation in the context of the development of the Solar System and beyond that in a galactic context, it has been realized that conditions and environments in the early history of the Earth were very different from those of the present. The present is not always a key to the past.

How to use the timelines in this book

Geological time has a very complicated structure, with many subdivisions and sub-subdivisions. This has evolved as geologists have discovered more about the age of the Earth and major events in its history.

In this book, we are using four main levels, or divisions of time. They are the eon (the biggest), followed by the era, then the period, and finally the epoch (the smallest); that is, several epochs go together to form a period, several periods form an era, and several eras comprise an eon. We have devised timelines to take you through these time spans in easy stages.

Each chapter has a double timeline on its opening spread. The top part shows the eons, and the eras within those eons, of all time in Earth's history. This top section is the same on every chapter's timeline. Part of this top line is then expanded in the lower section to show the tranche of time being discussed in that particular chapter. This lower line therefore shows these selected eras, and the periods within them.

Each chapter is comprised of several separate topics, listed in the Contents. Where these topics cover only a section of the time covered by the chapter overall, they are given their own single timeline. This shows the period or periods (which you will have already seen on the bottom section of the chapter timeline) relevant to that topic, and the epochs within them.

Chapter 9 is slightly different, since it talks about Earth's future. Thus the bottom section of this chapter's timeline travels millions of years into the future, where we do not know what any new eras or periods might be called.

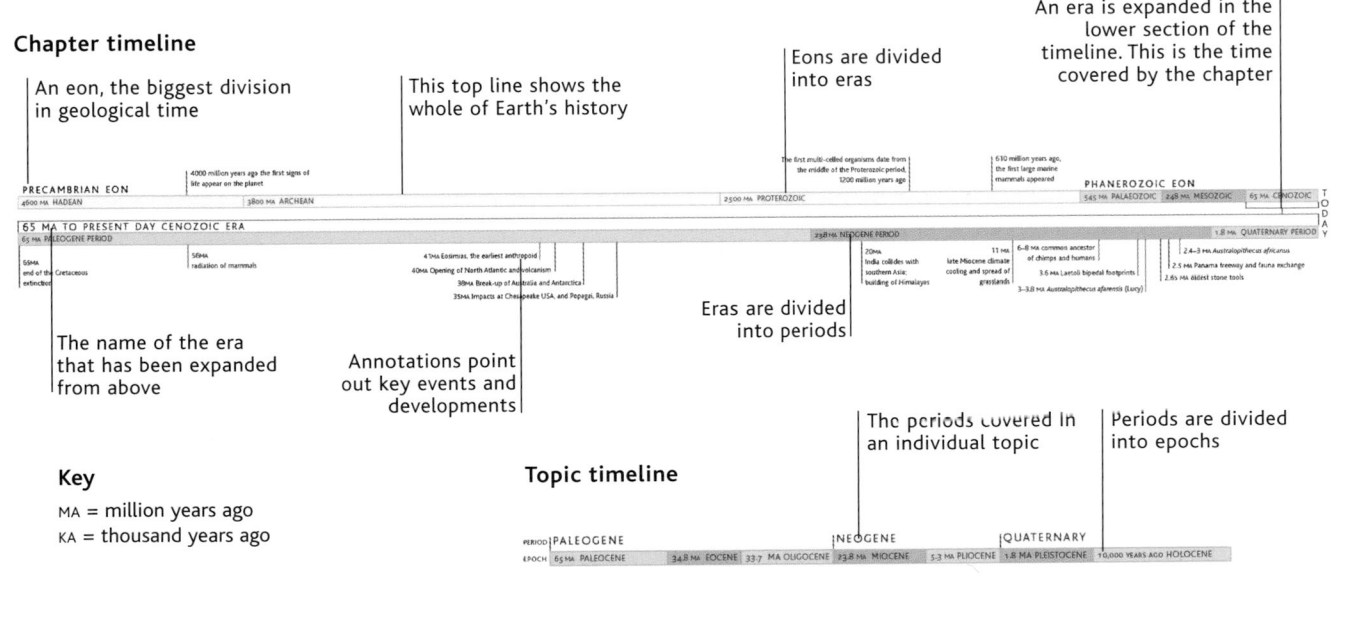

Chapter timeline

An eon, the biggest division in geological time

This top line shows the whole of Earth's history

Eons are divided into eras

An era is expanded in the lower section of the timeline. This is the time covered by the chapter

The name of the era that has been expanded from above

Annotations point out key events and developments

Eras are divided into periods

Key

MA = million years ago
KA = thousand years ago

Topic timeline

The periods covered in an individual topic

Periods are divided into epochs

1

How the present reveals the past

Nowadays it seems perfectly reasonable to study present phenomena in order to understand and interpret the past. Observing and measuring volcanoes, earthquakes, glaciation, climate change, weathering, erosion, and sedimentation, and studying their products is an obvious precursor to looking at rocks in order to work out how they were made. Similarly, the study of the biology of living organisms and their relationship with their living environments, known as ecology, can show how fossils fit into the natural world of the past.

However, this approach only became central to the practice of modern geology in the early 1800s, when taken up by Charles Lyell (1797–1875) in his highly influential book *Principles of Geology* (1830–3). The method was pioneered by the Scottish geologist James Hutton (1726–97), and the French anatomist Georges Cuvier (1769–1832) who took similar approaches but came from very different intellectual and cultural backgrounds, with Hutton a product of the Scottish Enlightenment and Cuvier a survivor of the French Revolution who espoused catastrophism.

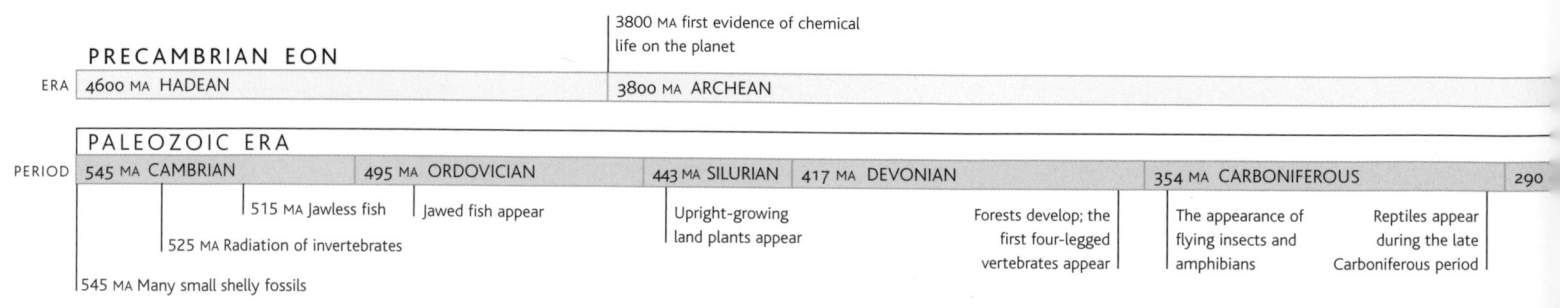

Cuvier studied the anatomy of different animal species alive today to see how body structures are modified for particular modes of life. From this he was able to work out the form and habits of extinct organisms even from only partial fossil remains. However, this method had its limitations; when Cuvier first saw the teeth of an *Iguanodon* dinosaur he thought they were rhinoceros teeth. Unsurprisingly, he could not predict the exact form of a completely extinct group of animals just from their teeth; nobody could.

Similarly, many rocks were difficult to interpret because the processes which form them, such as the deep intrusion of granite could not be directly observed. Also many geological processes cannot be observed in real time. Even more problematic were indications that events and environmental conditions were very different in the past. We now realize that there are events such as large scale impacts and glaciations which have not happened recently. Furthermore, the scale and frequency of many events are only now being fully appreciated.

Prehistorical records

Discovered in 1991, about 130m (426.5ft) below present sealevel off France's Mediterranean coast, the Cosquer cave contain an amazing array of Palaeolithic art, dated at 27,000 and 18,500 years ago. Cave art, such as this wonderfully observed horse provides evidence about Ice Age animal anatomy, not available from their fossil remains.

1200 MA The first multi-celled organisms date from the middle of the Proterozoic period

610 MA The first large marine animals appear

PHANEROZOIC EON

| 2500 MA PROTEROZOIC | 545 MA PALEOZOIC | 248 MA MESOZOIC | 65 MA CENOZOIC |

TODAY

MESOZOIC ERA

CENOZOIC ERA

| RMIAN | 248 MA TRIASSIC | 205 MA JURASSIC | 142 MA CRETACEOUS | 65 MA PALEOGENE | 23.8 MA NEOGENE | 1.8 MA QUATERNARY |

The first dinosaurs and early mammals appear

Birds and flowering plants appear in the late Jurassic period

Primates and songbirds appear

7 MA First hominids appear on Earth

The diversity of life today

Exploration of the Earth over the past few hundred years has gradually revealed an extraordinary abundance and diversity of life: from microbes living hundreds of metres below ground to coral reefs so large that they are visible from space. It also turns out that this diversity is far from equally distributed around the globe. Important diversity "hotspots" have been identified, especially in the tropics with their rainforests and coral seas. The same historical period has seen the exploration of the rocks and geological formations of the Earth. As successively deeper layers of strata were investigated, fossils were uncovered, revealing that the Earth and its diverse life have an extraordinary history, one that extends far back into the depths of geological time. So what does that history reveal about the origins of our planet and its multifarious inhabitants ranging from microbes to humankind?

The scientific understanding developed over the past few hundred years has revolutionized our view of life, its history and its origins. In the middle of the 18th century, only a few thousand different kinds of living organisms were known. From originally thinking that the Earth and its life were created over a short period of time in the not too distant past, scientists now have compelling evidence that the Earth is some 4600 million years old and that life probably originated as long as 4000 million years ago. Over this remarkably prolonged time span, life has evolved, increased, and diversified hugely, only to be dramatically and drastically cut back and yet recover. Perhaps surprisingly, it is still unclear just how abundant life on Earth is. And many of the details of life's history remain buried deep within the strata of the Earth awaiting the attention of future generations of scientific investigators.

Today, life occupies almost all niches – from deep below ground to the permanently frozen peaks of high mountains, from the depths of the oceans to high in the atmosphere. But this has not always been the case. At its very beginnings, life was confined to water for what was actually a very long period, and the colonization of land and air took some time.

Life's abundance and diversity today

J. B. S. Haldane (1892–1964) once remarked that, if there were a creator, then he or she "must have been inordinately fond of beetles". Despite all the efforts of scientists over the past few centuries, only about 1.7 million of the species of organism alive today have been properly described. Current estimates for the grand total of species of living organisms range between a very conservative eight million and a rather optimistic 100 million, with 30 million seeming the most reasonable estimate. But our understanding of the microbial world of bacteria and more primitive organisms is only just beginning. We may yet be underestimating the total by an order of magnitude and 100 million may indeed be more accurate. Perhaps the arthropods will yet be displaced from their dominant position as the biggest and most diverse group of organisms on Earth.

Both Alfred Russel Wallace (1823–1913) and Charles Darwin (1809–82) were profoundly impressed by the profligacy of life. They were also fascinated by the realization that, if all the offspring of almost any organism were to survive, within a few generations they would vastly outnumber all other creatures around them, dominate their living environment, and run out of

Arthropods

Of all the living organisms known today, a remarkable number are arthropods, especially insects. Arthropods have segmented bodies, paired and jointed limbs, and a toughened organic "skin" known as a cuticle. It has recently been estimated that there are between four and six million arthropod species alive today, among which perhaps half a million are beetles. By comparison, there are a mere 9000 species of bird, which in turn outnumber the 4000 or so mammal species. While these comparisons are quite interesting, however, they are not particularly useful because we are not comparing similar classificatory categories. Only those arthropods with mineralized skeletons such as the extinct trilobites, and crabs are well preserved in the fossil record hence our knowledge of ancient arthropod diversity is limited.

food. The notions of competition, adaptation, selection, and continuation of species were born independently in the minds of these two scientists. By a lucky coincidence of history, they came together in the mid-19th century when the Darwin and Wallace theory of evolution was first outlined in 1858. It was a British clergyman, the Reverend Thomas Malthus (1766–1834), who in 1798 first spelled out the problems associated with exponential and ultimately unsustainable growth resulting from high success rates in the reproduction of any organism, and these problems are still very much with us.

We have gradually come to realize through our growing understanding of Earth's history, its environments and inhabitants, and the processes that have sustained them that we live on a relatively small planet, the resources of which are not inexhaustible. The Earth's geological history shows that the progress of life through time has not been smooth, but rather filled with vicissitudes, booms, and busts, which at times have drastically cut life back and radically altered the face of the Earth. The driving forces for these changes lie both within and without the Earth and are mostly beyond our control. The best we can do is to learn as much as we can about them in order to better predict their recurrence, take avoiding action where possible, or, if avoidance is not possible, work out how best to cope with the impacts and "mop up" afterwards.

Coral reef
Tropical coral reefs have been the marine equivalents of the terrestrial rain forests for hundreds of millions of years. They support a remarkable diversity of life from simple algae and sponges to vertebrates, such as fish and turtles, many of whose remains can be fossilized, especially those with rock forming skeletons such as corals.

Rainforest

For over 350 million years, from late Devonian times, land plants have formed dense and diverse stands of forest, especially in humid tropical latitudes. Competition amongst the plants for light leads to different growth forms which provide food and shelter for a great variety of animals from microbes to insects and vertebrates.

The problematic record of "deep" history

It took a long time for scientists finally to accept what fossils are: the remains of once living organisms. There were very real problems of interpretation at first. So many fossils are preserved in strange ways with fossilization tending to obscure or remove all traces of original organic materials. Often, fossils retain only a physical resemblance to organic life, being preserved by purely inorganic and often crystalline materials. Even when fossils were recognized as being similar to living creatures, the presence of fossils in rock strata far inland and at the top of high mountains required an explanation.

To begin with, the only possible explanation seemed to be the Flood as described in the Old Testament. Not until the beginning of the 19th century was there any real understanding of the processes of petrification whereby life could be turned to stone. It then took another 100 years and more before the scientific revolution provided satisfactory understanding of the processes which transform seabed sediments and their organic remains into fossiliferous strata which have been folded, faulted, and uplifted to form mountains. Furthermore, it is only in the past decade or so that the unifying plate tectonic theory has emerged to provide a general explanation of Earth processes. For the first time, we can explain how fossils such as those found in coal seams originating in tropical rainforest trees or coral reefs from tropical latitudes come to be incorporated into high-latitude mountains thousands of miles away from where they originated.

At the beginning of the 19th century, scientists were still struggling to come to terms with the collapse of the old paradigms for the history of life and its origins. It is worth briefly recalling just how far we have travelled scientifically over this short time. For nearly two millennia, the world view that prevailed within the scientific community of Europe and the Americas was that of the Judeo-Christian tradition based on a belief in the historical truthfulness of biblical texts with their story of Creation and the Flood. According to this view, life was purposefully designed and created by a benevolent God for the express purpose of supporting the existence of His special creation, humanity on Earth.

Many of the scientists investigating the natural world were firm believers in this tradition, and many were ordained priests who justified their scientific work on the grounds that it would reveal the wondrous details of the Creator's work. But problems soon emerged. For instance, how could a benevolent God allow any of His creations to die out as the fossil record seemed to indicate? Nevertheless, to begin with, the fossil evidence did seem to support the biblical Flood story. By mid-19th century, however, the Ice Age theory had replaced that of the Flood interpretation. Even so, it was not until the end of the 19th century that human antiquity was generally accepted as scientific fact. Some people still do not accept it today.

Scientific investigation of the history of life on Earth, as recorded by fossils, has revealed that life has been abundant for a very long time – hundreds of millions of years, in fact. The further back in geological time we go, however, the less diverse it must have been. We now know that the environments of Earth were not colonized all at once. Life probably evolved in seawater and could not move into freshwater, onto land, then into the air until it was equipped to do so. As new evolutionary adaptations arose, life invaded new ecological territories or "spaces". Increasing populations and separation of populations promoted speciation, or the evolution of different species, with large jumps in overall diversity.

Unfortunately, it is very difficult to recover complete information about these increases in diversity because

of the nature of the fossil record. Fossils provide a view of the past that is heavily biased towards particular environments, mainly shallow-water marine ones. Fortunately, there are occasional "windows" where exceptional preservation gives privileged insights into the deep past. Fossil amber, the mineral replacement of soft animal and plant tissues, dehydration, and deep freezing – all have played a role in generating some of these windows.

So far, the known fossil record only encompasses some 200,000 or so ancient species for the whole of "fossiliferous" time. This represents an incredibly small sample of the totality of life. Suppose the average diversity of life has been 10 million species over the past 500 million years and that, with evolutionary turnover, species duration is 10 million years. This would mean that over 500 million years there have been 50 turnovers of 10 million species, producing a grand total of some 500 million fossil species. At present, we only know of some 200,000 fossil species, which is far fewer than 1 per cent and, in fact, only 0.04 per cent of that grand total. Even though palaeontologists still have a large amount of work to do, we will never be able to recover anything like the original total.

Global growth potential
Global biosphere data shows the distribution and abundance of oceanic phytoplankton at the base of the marine food chain (red and yellow representing high concentrations). The terrestrial measurements show potential for production of vegetation with green showing high growth areas and the buff colour representing low growth potential.

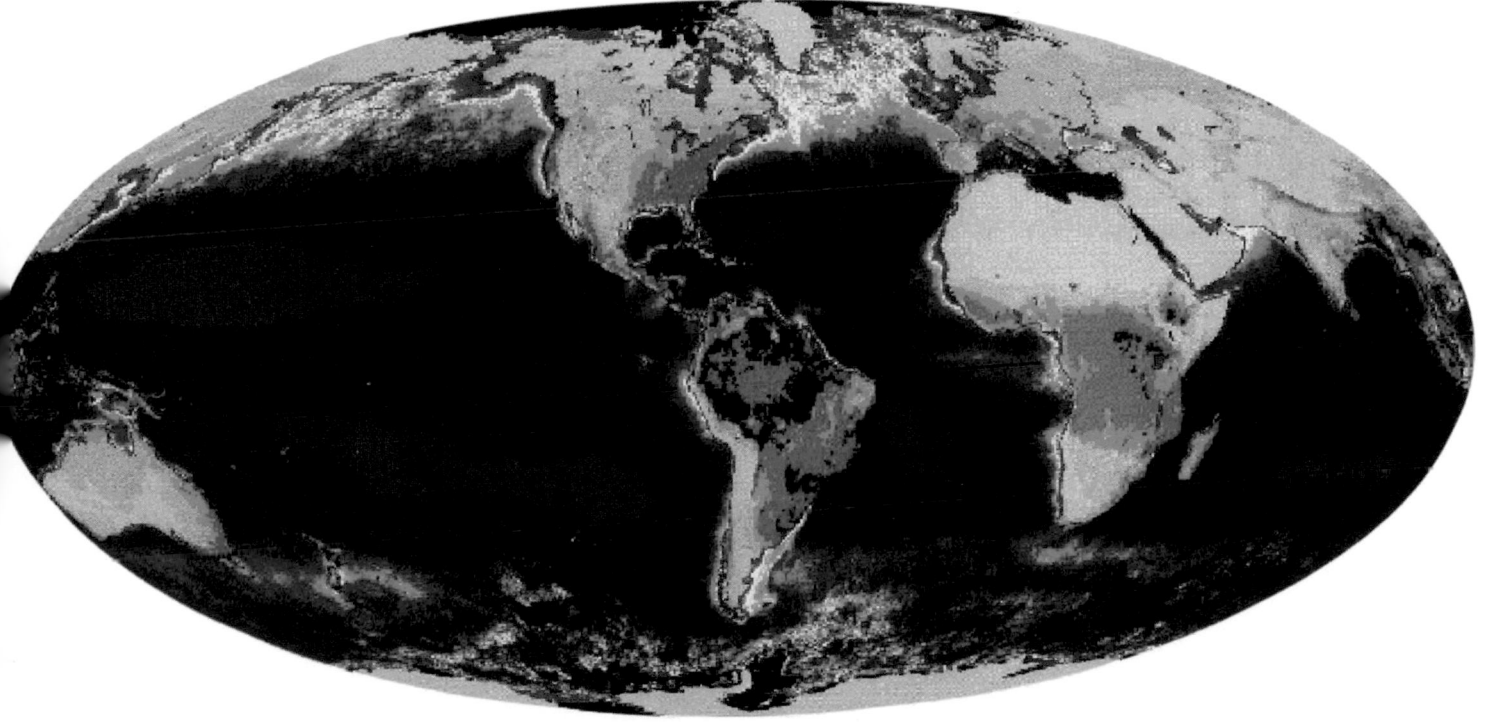

Life's relationship with the environment

Leopard

Natural selection has led to the survival of a remarkable diversity of body forms and lifestyles which are often repeated within different groups of animals and plants. Fleet-footed plant eaters, such as this small gazelle, preyed upon by fast running predators, such as this leopard, have coevolved over hundreds of millions of years.

All organisms are to some extent controlled by their surroundings. Most life can only exist within fairly narrow ranges of temperature, wetness/dryness, supplies of air and preferred food. As a result, life tends to be tied to particular environments. Polar bears and arctic foxes are well adapted to surviving in extreme cold as long as they have enough high-protein food. The African elephant can survive hot, semiarid desert conditions provided it can reach a water supply every few days and find enough plant material to eat in the meantime. They could not, however, swap places and survive for long. Yet, these three animals are all warm-blooded mammals and share a common ancestor who lived some 70 or 80 million years ago. Since that common ancestor roamed the Earth, these mammals have evolved and adapted to survive in very different climates.

If we only had the skeletal fossil remains of these animals to provide us with information, we could not determine the conditions under which they lived. As we know, bears, foxes, and elephants are all big groups of animals that include many different kinds (both at the genus and species level) which are separately adapted to different living conditions without necessarily leaving any trace of these differences in their skeletons. Still, all is not lost. Fossils are found in sedimentary rocks, and most sedimentary strata retain characteristics determined by the environment in which they developed and were deposited. By analysing the sedimentary context and accompanying fossils, especially plant remains, scientists can often deduce the kind of environment in which the animals and plants of the past lived.

The geographical distribution of different life forms varies enormously. Many parasitic organisms depend on just one or two host species for survival, so their geographical range is intimately tied to that of the host. Many flowering plants depend upon certain insects for pollination, and many animals depend on just one or two plant species for food. These are all examples of specialists, the distribution patterns of which are linked. By comparison, generalists such as humans, foxes, and

bears can cope with a large range of food types and consequently can respond quite well to sudden changes in food supply. But humans have not always been so adaptable. It is likely that several human-related species, such as *Paranthropus boisei*, became extinct because they were too specialized in their feeding habits.

The close interrelationship between organisms and their environments and constraining physical and biological parameters raises an interesting problem. How is it possible for life forms ever to leave the "home" environment, the one to which they have adapted, for a new and different one? There is good fossil and biological evidence to show that this has happened time and time again. For instance, several different groups of organisms have at times left the environment of the sea or freshwater for dry land, while others have literally taken flight from land. Life cannot simply wish such changes to happen in order to exploit new vacant plots or spaces with very different physical and chemical conditions. Getting about, breathing, and feeding in a supportive liquid medium such as water is very different from trying to do the same things in a light, dry gas such as air. Adaptation must begin even before such new niches can be exploited. How do such preadaptations come about?

As we are learning to our cost, the environments of the Earth are not as stable as was once thought. The Earth's surface is dynamic and constantly changing, even though the rates of change can be very slow. Changes can also take place at an uncomfortably rapid pace, as we are beginning to realize. Climate, in particular, is a major controlling factor on life. Most plants require certain levels of rainfall, temperature, etc. Consequently, patterns of vegetation distribution tend to be zoned according to climate patterns. Animal life ultimately depends on plants for food, and, if plant distribution patterns are disrupted by changing climate, there can be a cascading collapse through the food chain from the plants to the plant-eating animals and finally to the carnivorous ones.

Investigation of the geological record shows that the Earth's environments have changed drastically in the past and therefore so, too, must have past climates. In northeastern North America and northwestern Europe, the ancient rock succession around 380 million years ago (Ma) shows a sequence from semiarid desert conditions in Devonian times, through shallow seas full of coral reefs in Carboniferous times (c. 340 Ma), followed by rainforests, and back to hot, dry deserts in Permian times (c. 280 Ma). This entire rock sequence is now found in high latitudes that often experience winter snowfalls. When first discovered, these very contrasting data were seen as strong evidence that climates had changes drastically in the past. It is much more complicated than that, as it turns out, because the landmasses have not always occupied the geographical position that they now do.

Nevertheless, climate change certainly has happened – and in the not too distant past – with a series of climate oscillations which we now know as the Quaternary ice ages. These changes had a drastic impact on environments and the life which they support.

Polar bear
The extremes of polar climates have only really been successfully conquered by plants and animals (such as warm blooded mammals and birds) which can adapt to the sub-zero temperatures and limited food resources. Here, the well camouflaged arctic fox and polar bear are both predators that depend upon the superb insulation of their fur.

Processes which change environments over time

Our view of the Earth and the stability of its environments has been radically changed by the discoveries of geology and new techniques of observation. We now have a perspective on the past which spans the entire 4600 million years of Earth's history. Inevitably our view of the recent past is much more detailed than that of remote times. Nevertheless, some major events and episodes in Earth's early history which have impacted upon surface environments and their inhabitants, however primitive, are gradually becoming clearer. We now have good evidence that there were fundamental differences in the composition of the Earth's early atmosphere. Even the remote past was punctuated by ice ages and extra-terrestrial impact events which must have had severe effects on primitive life and its evolution. Oceans were opened and closed, with continental masses growing, breaking apart, and being shuffled about over the Earth's surface.

The study of the biological relationships of organisms to their living environments has shown that most life is adapted to tolerate and survive some change, whether from a twice-daily rise and fall in sea level, a temperature change from day to night, the lunar monthly tidal cycle, or seasonal or annual changes. Even if these familiar rhythms are disrupted or shift in any significant way, organisms have changed either by moving, adapting, and evolving, or by dying out.

With more than 150 years of modern scientific measurement and analysis of living environments behind us, many processes of change are now familiar and reasonably well understood, if not necessarily predictable. Processes which effect major changes on Earth's environments tend to operate on very different time scales, frequencies, and geographical impact from those which operate at local and annual to regional and decadal, or longer, time scales. We are most familiar with the types of changes that are frequent and widespread, such as climate fluctuations operating on an annual or decadal scale, and which can be experienced within an individual lifetime.

Powerful storms and flooding which have enough energy to damage property and even threaten lives are familiar events that occur annually in many parts of the world. Thanks to modern observation and analytical techniques, we have a better comprehension of how, in any one location, the magnitude and frequency of these weather phenomena may change as climate changes.

Similarly, earthquakes and volcanic eruptions are frequent in some regions of the world, especially around the Pacific rim from Patagonia to New Zealand. We build cities in earthquake zones and within the destructive shadow of active volcanoes because of the pressure for living space and fertile land. Again, life tends to accommodate to such processes, but the geological record shows us that repeatedly in the past much larger scale but rare events have happened and will happen again, especially volcanic eruptions. There are vast regions of India, Siberia and northwestern United States which are covered with great thicknesses of lavas known as plateaux or flood basalts that poured out of fissures in the ground and literally flooded surrounding landscapes, destroying life. As no such events have happened within recorded history, it has taken a long time for scientists to realize that they do happen.

Destructive tsunamis (tidal waves) can be generated offshore by earthquakes, submarine volcanic eruptions, and slumps of huge masses of seabed sediments. These

Hurricane Mitch

We now know that rare but exceptionally powerful storms and unusually extensive flooding can happen on decadal (10-year) and centennial (100-year) frequencies. Hurricane Mitch, which occurred in October 1998 and killed some 11,000 people, became known as the storm of the century, and it was certainly the worst to hit the Americas since the 1780 hurricane which probably killed more than 22,000 in the Caribbean region. The effect of such rare events on environments can be more devastating than all the intervening annual events.

events displace huge volumes of water locally and send shock waves across oceans to wreak havoc thousands of miles away. Until recent decades, most scientists (except those in Japan, which had direct experience of such events) had little knowledge of this phenomenon. We now suspect that tsunamis perhaps as much as a kilometre high have been generated by the impact of large, extra-terrestrial bodies in the oceans. Ancient sand layers, some 65 million years old, found around the Gulf of Mexico are thought to have been deposited by giant tsunamis generated by the Chicxulub event which marked the end of the Cretaceous period and the final demise of the dinosaurs. Again, fortunately, no such large-scale event has happened within recorded history. It is enough to know of the destruction to the Portuguese capital of Lisbon in 1755, and the deaths of 70,000 people, when an earthquake hit the city and also generated a tsunami that compounded the destruction.

Devastating landslides were once thought of as relatively rare events. We now know, however, that they are common in geologically young mountain ranges such as the Himalayas and the Andes, and even in less dramatically mountainous regions such as California. Landslides can be generated by earthquakes or heavy

Sand dunes

Terrestrial landscapes are normally being worn down to sealevel, under the influence of gravity by the processes of weathering and erosion. However, there are also some important environments such as deserts where there is a net accumulation of sediment which stands a considerable chance of being preserved in the future rock record.

Changing climate

Satellite observation has recently shown significant changes in and the break-up of major ice sheets, such as that of the Larsen Ice Shelf, the largest on the east coast of the Antarctic Peninsula, due to global warming. The exact effect of the release of such large quantities of cold ice on the life of the surrounding seas is not yet known.

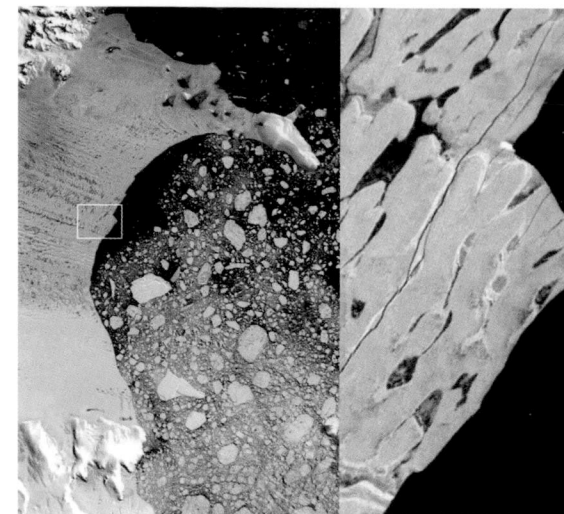

rainfall, and they are common events wherever this combination occurs in seismically active, hilly terrains such as the monsoon-drenched regions of Southeast Asia. These processes have a considerable impact on the environment and its inhabitants. Under the influence of gravity, rock and soil debris avalanche downslope into streams and rivers, which carry the debris away and dump much of the load on their downstream flood plains. The hillside scars expose fresh rock to the slower processes of weathering, soil formation, and invasion by vegetation before the whole cycle happens again. The net effect is constant turnover and transport of Earth materials, part of an overall recycling. But if climate change were to take place in the form of, for example, cessation of the monsoon, there would not be enough rain to support such profuse vegetation and soil development, and the environment would not be able to support so much life.

Just how dramatic the environmental effects of climate change can be is best seen around the margins of arid regions such as the Sahel in North Africa. The occasional rains and sparse scrub vegetation of semiarid regions are often just about enough to support migrant populations of game and people with herds of domesticated cattle. If those sparse rains fail, however, even the desert-adapted plants will die or be used up by desperate people, with the result that the environment becomes uninhabitable. Without plants, the thin soils deteriorate, and the sediment is more prone to wind erosion and the encroachment of shifting dunes. Interestingly, there are recent reports suggesting that some of the desertification has been reversed.

We humans have notoriously not been good "stewards" of the land. In the past, much of this has been through ignorance of the "knock-on" effects. The overexploitation of the prairie soils of North America's Midwest in the early decades of the 20th century is a good example of this, with increasing mechanization allowing vast acreages to be ploughed for planting grain. Hot, dry summers and soils unprotected by the native grasses led to wind erosion, the development of "dust bowls", and eventually badlands. We have no such

Extinct birds

One of the most fascinating images of the submerged Cosquer cave is this charcoal outline of a bird, first thought to be that of a penguin but now recognized as an accurate image of the extinct penguin-like flightless great auk which was finally hunted to extinction in the mid 19th century. Fossil remains of the great auk have been found around the Mediterranean whilst penguins are inhabitants of the southern hemisphere of the globe.

excuses today because the vulnerability of plant cover, soils, and whole environments is now well understood; however, population pressure continues to have a deleterious effect in many regions. Virgin forest is still being cleared, first to provide timber, then space for agriculture. Not only is there a serious loss of habitats for many endangered species, but also these forests are thought to act as carbon dioxide "sinks" which help to ameliorate the effects of global warming.

Sea-level changes were a constant feature of the distant past and can result from a number of geological processes. The best known is the kind of drastic climate change associated with ice ages. So much ice is produced that significant quantities of water derived from the oceans are locked up as ice on land, and global sea levels drop. Global warming heats oceanic waters sufficiently to cause them to expand, adding to rising sea levels. When the ice eventually melts, the process is reversed, but the slow depression and rebound of the ice-laden land complicate the net effects. For instance, at present Northern Europe is still rising due to rebound after the melting of the Quaternary ice sheets, despite the fact that sea levels are rising due to global warming. Land movements on a local and regional scale can also produce relative movements of land to sea level. Finally, the relative rates of ocean floor spreading can affect global sea levels by changing the volume of ocean basins.

Falling sea levels create land bridges such as that between Asia and North America, which allowed the migration of animals, including humans in the prehistoric past, between the two great continents. Rising sea levels isolate landmasses and their inhabitants, from small islands to continents. The repeated isolation of North America from Asia has had a huge effect on the migration, population isolation, and subsequent evolution of mammals within these continents. Britain and Europe, and Australia and Southeast Asia have similarly been isolated by rising sea levels, which also led to the loss of important productive lands marginal to the sea around the world. Shallow continental-shelf seas flood into low-lying continents and can in turn become highly productive in terms of marine life. The interplay of rising and falling sea levels has been a very important process in the geological past, eg in the development of the coal measure forests of Carboniferous times.

One major process which affects environments in the long term is that of plate tectonic movement, which has opened and closed oceans, moved continents and broken them into pieces, only to reassemble them in new patterns. This process sees margins of colliding plates being crumpled together, thickened by tens of kilometres, and folded and faulted to form major mountain chains such as the Himalayas. The migration of these plates has carried some life along with it, but it has also destroyed an enormous amount of evidence about life in the oceans of the past.

Volcano
Although generally perceived as destructive, volcanoes are often constructive geologically. Their rock products (airborn rock fragments, magma, ash and lavas) often cover and preserve surrounding landscapes and sometimes the remains of animals and plants. When cooled and lithified such volcanic rocks often prove extremely resistant to weathering and erosion.

Earth— a dynamic system in constant flux

From space, the Earth is a blue planet with plenty of water shrouded by a patchy, white cloud cover showing the presence of a humid atmosphere, below which landmasses coloured green, brown, and yellow can be glimpsed. By stark contrast, Earth's satellite, the Moon, is a barren and heavily pockmarked sphere of rock and dust with no signs of life. And yet we now know that the Moon is some 4000 million years old, only 600 million years younger than the Earth. So why do these two bodies look so different? Why is the Earth not just as craterous as the Moon?

Indeed, all planetary bodies are subjected to bombardment by objects of all sizes ranging from a daily rain of dust to large asteroid sized rock masses every few million years. The planetary bodies in the Earth's solar system are also subjected to temperature change because of their orbits, axis of rotation, and rate of spin relative to the Sun. Those with gaseous atmospheres have winds as well, and, if there is any water in the system, there is the possibility of ice and ice caps, steam, clouds, rain, and oceans. The state of the moisture depends, however, on the ambient temperature, so that Europa, one of Jupiter's moons, has a surface temperature of minus 170 degrees and is covered in ice. But it is evident that this ice cover is remodelled in a cycle which starts again every 10 million years or so.

If the Earth were "dead" as the Moon is now, its surface would look like that of the Moon, frozen in time and still pockmarked with craters from ancient impacts and volcanism that produced vast ancient lava "seas". The Moon has no internal heat energy, no atmosphere, and no water. Although the Earth has been subject to events similar to the ones the Moon has, the constant recycling of the Earth's surface has destroyed much of the evidence. Luckily for us, the Earth is still very much "alive" and dynamic, otherwise there would be no life on our planet. There are several reasons for the Earth's continuing vitality. Its distance from the Sun, its orbit, its axis of rotation, and its rate of spin – all help to produce a relatively equitable surface environment. It is neither too hot to burn off all the moisture nor too cold to freeze it completely. There is enough water to generate an atmosphere and oceans, which further help to cushion us from the extremes of solar radiation and their potentially harmful rays (especially ultraviolet light).

Most importantly, our planet has an internal source of heat that drives the great Earth machine with its continual convection heat flow mechanism. This ongoing mechanism creates and destroys oceans and drives the movement of the vast crustal plates into which the surface of the Earth is fragmented. The action and processes of convection currents and plate movements are large scale and long term. While the mechanisms are still far from being fully understood, there is incontrovertible evidence that plates can move over thousands of kilometres over tens of millions of years. This may seem extraordinary, but spreading rates of 10cm (4in) per year are equivalent to a kilometre every 10,000 years and 100km in a million years. Subducted slabs are recycled within the Earth over even longer periods – probably hundreds of millions of years.

Overall, the effect has been to give the Earth a complex and dynamic history which geologists are still trying to unravel. The most recent geological past is the most familiar; the further back in time we proceed, the less clear the story becomes. The past 200 years of modern scientific investigation have given us a fairly good idea of the outlines of Earth history back over the past 500 million years or so. Beyond that, the much greater stretch of Precambrian time is still poorly known, but is the subject of active investigation.

A general look at the main topographical features of the Earth's surface provides clues to the patterns and mechanisms of our planet's internal heat machine. In comparison to our human scale, variations in the Earth's surface topography seem to be extreme. Mountains rise so high above sea level that temperatures are permanently below freezing and oxygen is in short supply at their peaks. (Everest, for instance, is nearly

9km (6 miles), or 8846m (29,022ft), high.) They are so inhospitable to life that humans can visit them only briefly; otherwise they are normally devoid of life. These high mountains typically occur in elongate ranges extending over hundreds or thousands of kilometres. Geological exploration of them has shown that each mountain belt has a particular history related to very large-scale processes. Their rocks are folded and faulted by massive compression, thickening the crust to such an extent that their "roots" have partially melted, producing volcanoes and large intrusive rock bodies.

The global distribution of major earthquake zones with their related faults, volcanoes, and eruptive products all provide clues as to underlying dynamic processes within the Earth. Geologists study these phenomena today and relate their findings to evidence for the same processes which occurred in the past. As a result, the "deep" geological history of the Earth has been mapped out and is now quite well established for the past 500 or so million years, and this history is gradually being extended deeper and deeper into the more ancient past.

Hurricane force

Cyclonic storms, such as Hurricane Andrew seen here in the Gulf of Mexico in August 1992, are normal seasonal events in many tropical regions. In recent decades it has become evident that every hundred years or so superstorms of even greater magnitude can occur, which have the potential to leave lasting marks on the rock record, especially in continental-shelf sea deposits.

Landslide

We humans tend to be rather shortsighted, and persistently underestimate the power of the natural processes of weathering and erosion. Hillslopes and coastal cliffs might seem perfectly strong and longlasting when dry. However, in the long term they invariably crumble away under the combined effects of heavy rains, slumping, sliding and avalanching of rock and sediment.

The greatest revolution in the understanding of the Earth's dynamism has resulted from exploration of the oceans over the past few decades. While details of the Earth's land topography have been known for many years now and carefully mapped by surveyors, mapping of the bottom topography of the oceans (some 70 per cent of the Earth's surface) has been a much more recent achievement. It has depended upon sophisticated ship-borne techniques and more recently the same type of remote satellite sensing that has been applied to the mapping of the land, its topography, soils, vegetation, and structural features.

Major topographical features have been discovered which tell us a great deal about the inner workings of the Earth. Some of the most dramatic topographical features are the ocean deeps such as the Marianas Trench, which extends 11,033m (36,198ft) below sea level. There has to be some deep-seated "reason" for the ocean floor suddenly to descend to such depths, some force has to be pushing it down, a force that can now be explained by plate tectonics.

The difference between these depths and the highest mountains gives an overall vertical range of nearly 20km (12 miles). Yet most of the land surface has no great elevation. Indeed, the average height of the land is a mere 840m (2757ft) above sea level, while the average depth of the oceans and seas is 3.8km (12,460ft) below sea level. Not only are the oceans significantly deeper than the land is high, relative to sea level, but also the area of the Earth surface taken up by the oceans is much greater than that of the land (70 per cent). The oceans dominate the Earth's surface and have done so for billions of years. Moreover, most of the history of life has been in the oceans. Life only set "foot" on land for the first time around 460 million years ago.

This variance in average height of the land and the depth of the oceans is caused by an underlying fundamental difference in the rock materials from which they are made and has a far-reaching implication in major geological processes. Basically, the oceans are floored by igneous rocks that are denser than the rocks which make up the continents. The ocean floor is not just marked by deep ocean trenches. It also features the longest mountain chains on Earth. One of these oceanic

mountain chains runs southwards in mid-ocean from Iceland down the length of the whole Atlantic, before diverging with one branch extending eastwards into the Indian Ocean and the other extending westwards to the southern tip of South America. Compared with the more familiar mountain ranges on land, these submarine ranges have some interesting and important differences. Not only are the ranges much longer, but they are symmetrical in cross-section as well, and it is no accident that they are known as mid-ocean ridges because they are also spreading ridges which mark the site of the ocean growth. Why are there mid-ocean mountain ranges, and why are they different from mountain ranges on land? Again, plate tectonic theory is providing an explanation.

Other major features of the ocean floor include chains of volcanic ocean islands such as the Hawaii–Emperor chain which extends thousands of kilometres from mid-Pacific northwest to Kamchatka. The islands of Hawaii are one of the biggest and most active volcanic complexes on Earth, in that they rise some 5000m (16,400ft) from the ocean floor, then a further 4000m (13,000ft) above sea level. Northwest from Hawaii, the islands of the same chain become less active, less high and geologically older. Eventually they do not even rise above sea level, but continue as a line of sea mounts that sank beneath the waves as they cooled. Their existence and nature are signs of an ocean crust plate moving over an active "hot spot" deep within the Earth. As long as the Earth is a dynamic system in constant flux, such processes and events will continue. They may be difficult to live with, but without them the Earth would "die".

Isostasy

Under the principle of isostasy continental masses "ride" high on the denser rock of the ocean floor and the semiplastic rocks of the Earth's interior (the mantle) below, similar to an iceberg floating in the sea. As icebergs melt, or continents are worn down, they continue to rise, maintaining the same position relative to sealevel as the material they displaced returns beneath them. Icebergs float with about 10 per cent of their mass above water. Similarly, continents, which have an average thickness of around 35km (21¾ miles), show only a small proportion above sealevel.

Environments today as potential sites of fossilization

If you have a particular desire for immortality and fossilization of your mortal remains for hundreds of millions of years, then you have something of a problem. Burial on land can work. After all, the fossil record of dinosaurs largely depends upon it, but the vast majority of fossils are of marine organisms. To be preserved as fossils, organic remains generally need some preservable hard parts and have to be buried deep within the Earth. Unfortunately, landscapes do not normally make good long-term burial sites because they are continually eroded, with their rock debris eventually being dumped at sea by rivers or wind.

The vast offshore sites of sediment accumulation which form the shallow water shelves around the continents are the main burial grounds for potential fossils. These layers of sediment build up into piles many hundreds or even thousands of metres thick and over time become compacted into sedimentary strata. Burial in the deep oceans is as problematic as burial on land, however, because most ocean floor is eventually subducted by the processes of plate tectonics. But the thick piles of sea-shelf sediment fringing the continents do not subduct so easily. Rather, plate collision deforms and uplifts these vast aprons of sediment to form new mountain ranges for future geologists to hammer away at, looking for fossil evidence to find out their age and environments of deposition.

So, for any remnants of ancient environments to be preserved in the rock record, various conditions have to be met and a succession of events has to take place in the right order. In general, the right combination and sequence are not found as often as one might expect. If we pause to consider the dynamic nature of the most common of Earth's surface environments, however, it soon becomes clear why this is so. Landscapes are for the most part above sea level. All the processes of weathering and erosion impact upon them. Under the influence of gravity, any loosened surface material, whether carried away by water or wind, essentially moves down slope and is eventually carried to sea level.

This overall trend may take a very long time – many millions of years in some instances – but it is fairly inexorable. In the long term, landscapes are worn down by this process of peneplanation, as it is called, to a base level close to sea level. Even the highest mountains can be worn down and removed, with their debris dumped in the sea, over a very long period of time. The fact that we can see igneous rock such as granites, which were originally intruded at depths of tens of kilometres and are now exposed at the surface, tells us that tens of kilometres of surface rock material can, with sufficient time, be stripped away. However, there are important geological exceptions to the general trend of removing sediment from land and dumping it at sea. The most important is the formation of sediment traps within continents, either within topographic or structural basins.

Large topographic basins of sediment can form where high rates of sediment accumulation combine with subsidence of the crust over a long period of time. Rapid sediment deposition tends to occur on the flanks of rising mountain chains where the crust is stretched and thinned, and tends to sag. High rates of erosion and deposition lead to the development of large alluvial fans, river flood plains, and, sometimes, lakes or inland seas as well. The accumulating weight of sediment further depresses the crust and allows thousands of metres of deposits and their contained fossils to build up. Such depositional environments develop on the flanks of young mountain belts such as the Alps and Himalayas. Rapid uplift of the mountains promotes increased rates of erosion (with land surfaces being reduced by over 6mm (¼in) per thousand years), transport and deposition of sediment on the surrounding lowlands. For instance, the Ganges–Brahmaputra region south of the Himalayas is a site of vast sediment accumulation with over 1500 tonnes per square kilometre deposited each year. Such thick sequences of continental strata have a good chance of being preserved in the rock record, as do those that accumulate in rift valleys.

From the human point of view, the role that rift valleys have played in preserving strata and fossils is very important. This is due to one rift valley system in particular, the Great Rift Valley of East Africa, a remarkable feature so large that it can be seen from space. The valley stretches some 8000km (5000 miles)

from Mozambique northwards through Kenya to Ethiopia, the Red Sea, and right up into the Jordan Valley.

Rift valleys originate where the Earth's crust is being stretched. As rocks are mostly brittle, rather than elastic, they tend to break along fault lines. Between parallel faults, the intervening rocks sag or collapse like the keystone of an arch that fails through lack of side supports. The downfaulted central valley then becomes a site where sediment can accumulate over a long period of time and stand a good chance of being preserved even when the surrounding landscapes are worn down and removed by erosion and weathering. In the example of the Great Rift Valley, evolutionary, geological, and geographical accident or coincidence have luckily combined to preserve a record of much of what has happened over the past 15 million years or so. Both sedimentary environments and a remarkably good record of the life of that period can be found in these rocks. Through a fortunate piece of timing, this just happens to coincide with a critical phase in the evolution of early humans and our extinct relatives. If ever there were an Eden, this is where it was. Charles Darwin was right: a critical part of our history and evolution is to be found in Africa. Indeed, genetic studies have recently shown that all living humans are essentially African in ancestry.

Pompeii
The eruption of the southern Italian volcano of Vesuvius in AD 79 produced a number of catastrophic flows of hot gas and ash, known as pyroclastic flows, which rolled down the mountain overcoming the towns of Herculaneum and Pompeii, suffocating many of the citizens. Such catastrophic events which preserve an instant in time are known today as "Pompeii events".

Underwater plane wreck (far left)
Examination of the successive colonization of submerged and accurately dated historic wrecks gives useful measures of the rates at which marine organisms grow and at which remains are buried within seabed sediments.

Our view of the geological past

Grand Canyon
America's Grand Canyon is one of the few places on Earth where it is possible to see a sequence of strata stretching back over a 1000 million years from the Cenozoic era right back into the Precambrian eon.

We now know that the Earth has a lengthy history extending back around 4600 million years. Life on Earth also turns out to have had a remarkably lengthy history, with fossil evidence appearing in the rock record perhaps around 3800 million years ago and certainly by 3500 million years ago. Strangely, it also appears that it took a very long time for life to "get going" in its evolutionary cycle. Much of its early history was microbial, and it took a long time for multicellular life to evolve, perhaps as much as 1500 million years. Increase in size and biological complexity to a level where fossils became readily visible to the unaided eye took even longer. Not until late Precambrian times around 600 million years ago, some 3000 million years after life actually began, do we begin to find what most of us would think of as fossils. Even then, these creatures were all still bound to the seas, and at this stage were entirely soft-bodied. The evolution of hard shells and skeletons had not yet

begun. So life seems to have had this extraordinarily long "slow-burning fuse" within Precambrian times.

Part of the reason for the slowness of early evolution may have been the changes that took place in the early development of ocean and atmospheric chemistry. We know that there was something of a "chicken-and-egg" problem in the oxygenation of the early atmosphere. Degassing from early volcanic activity provided the first atmosphere, and it was dominated by carbon dioxide (about 80 per cent); initially, there was no free oxygen. Gradually, moisture in the atmosphere was cracked into its components, oxygen and hydrogen, resulting in the slow build-up of oxygen and the escape of the lighter hydrogen into space. Early sedimentary strata rich in chert, which forms mainly in low-oxygen environments, bear witness to this state of affairs. The slow oxidation of the atmosphere is also recorded in the accumulation of iron-rich sediments called banded iron formations which today provide most of the world's iron ore. It appears that it was not until around 1800 million years ago that oxygen appeared in any significant quantities in the atmosphere and allowed the evolution of organisms that "breathed" oxygen.

The subsequent history of life over the past 500 million years and more has been well known, at least in its broad outlines, since the early 1900s. The biologist Peter Medawar (1915–1987) thought that this outline was so well known that any further investigation of fossil life was merely a matter of "dotting the i's and crossing the t's". As we shall see, he was wrong, unless you consider the discovery of feathered dinosaurs as merely a case of "dotting an i".

Two hundred and fifty years of scientific exploration have shown us that we have a geological rock record accessible at the surface of the Earth. Surprisingly, this record extends back almost as far as the origin of our planet. It took most of the 19th century for geologists to carve up geological history into a succession of identifiable periods of time such as the Carboniferous and Cambrian. The efforts of explorers and surveyors who discovered new lands, lakes, and mountains are often celebrated and well known. Yet who knows of the Reverend Dr W. D. Conybeare and his colleague W. Phillips who named the Carboniferous period in 1822, or of the Reverend Professor Adam Sedgwick who named the Cambrian period in 1835?

This mapping of the temporal dimension of Earth history through its rocks was a remarkable achievement, one that few people fully appreciate. Long before geologists even believed in evolution, they realized that the fossil content of successive strata changed and that fossils could be used to distinguish strata. The complication is that many strata are folded and faulted by Earth movements and that there are large areas of the land surface where no rocks are exposed. Nowhere on Earth do we find anything like a complete sequence of rock strata representing all the different phases of Earth history. Only by a long and complex procedure of mapping and dating rock strata, and comparing their sediments and fossil content from different sequences around the world can a composite log record of geological history be compiled. In any one location, the local sequence of strata will be full of gaps when no sediments were recruited to the record.

Darwin was right when he complained of the incompleteness of the geological record. In fact, the gaps in the record are often telling us about important events other than deposition, notably processes that have led to uplift and erosion, often associated with mountain building and related to plate tectonic movements. The investigation of the huge range of Precambrian time is one of the last great unknowns of geology. Much has been discovered — its life and the existence of major changes in environments over these 4000 million years, including astonishing ice ages — but so much more remains to be discovered.

Insect in amber

The preservation of past life as fossils varies enormously in its quality and completeness, from the occasional broken bone, shells, or footprints, to the almost perfect preservation of organisms trapped in amber resin. However, despite appearances and past claims, amber does not preserve ancient biomolecules such as DNA.

2
The ice age

The gradual realization that an Ice Age, rather than the biblical flood, scattered sediment and the fossils of large mammals across northern Europe and North America opened a new window on the prehistoric past. One kind of catastrophic event was replaced by another just as extraordinary and with far reaching consequences for humankind. There was mounting evidence for glacial activity as far south as New York, Birmingham and Berlin. A Swiss geologist, Louis Agassiz, was even arguing that a global glaciation might have reached as far south as the Amazon. How and why had it happened and what effect had there been on the life of the times? Were humans exposed to such drastic climate changes? The mid 19th century was a period of scientific turmoil with a constant deluge of new discoveries about the natural world and its prehistory. There were intense debates over the interpretation of the finds. Could they be fitted into a Lyellian vision of gradual change over enormously long periods of geological time, or was the older idea of the role of catastrophism making a comeback?

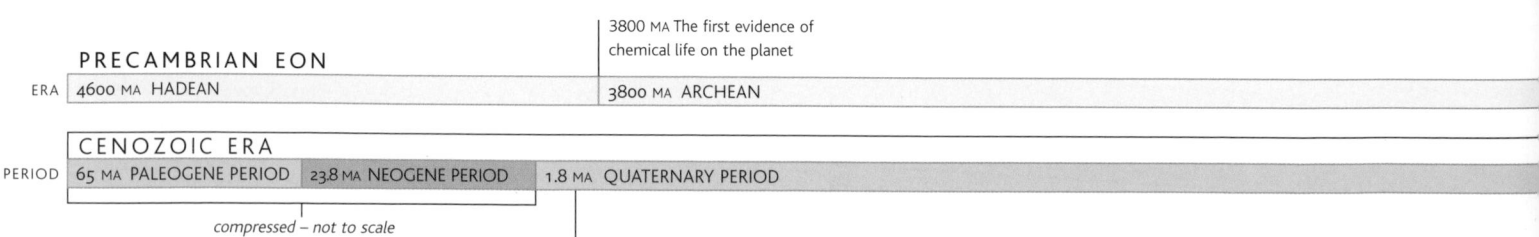

3800 MA The first evidence of
chemical life on the planet

PRECAMBRIAN EON

ERA | 4600 MA HADEAN

3800 MA ARCHEAN

CENOZOIC ERA

PERIOD | 65 MA PALEOGENE PERIOD | 23.8 MA NEOGENE PERIOD | 1.8 MA QUATERNARY PERIOD

compressed – not to scale

1.7 MA *Homo erectus* migrate out of Africa

The whole question of Ice Ages and what causes them has taken on considerable importance over the last decade or so. Climate change has become increasingly important as the full implications of what even a few degrees of temperature change over several decades will mean to life on Earth have become apparent, and widely publicized.

The acceptance that our human activity does and will continue to impact upon climate change has spurred investigation of the mechanisms involved.

Furthermore, there has been the sobering discovery that very rapid climate change has happened during the recent Ice Age and may well happen again in the not too distant future. Again the questions of how and why have these changes come about?

There is a considerable urgency today to understand the very complex global climate system with its interlinking of ocean and atmospheric circulation. In the 21st century, we are already witnessing the effects of gobal warming.

Greenland ice sheet

Ice is a remarkable material which can act as a solid but can also flow in a semi-plastic state. Here this ice sheet flows through gaps in mountains to spread onto lower ground. Such behaviour explains the power of glaciers and ice sheets to grow and cover huge areas in relatively short periods of time.

1200 MA The first multi-celled organisms date from the middle of the Proterozoic period

610 MA The first large marine animals appear

PHANEROZOIC EON

2500 MA PROTEROZOIC | 545 MA PALEOZOIC | 248 MA MESOZOIC | 65 MA CENOZOIC

TODAY

150 KA Penultimate glacial maximum
120 KA Last interglacial
100 KA Modern *Homo sapiens* migrate out of Africa
80 KA First bone tools. Graphic symbols in Africa
60 KA *Homo sapiens* arrive in Australia
c.20 KA *Homo sapiens* arrive in North America

30 KA Extinction of the Neanderthals
18 KA Last glacial maximum; lowered sea levels

12 KA Last retreat of ice caps and glaciers
10 KA Human-induced extinction of big game
7 KA The beginning of agriculture
4.8 KA Extinction of last mammoths
Little Ice Age 1500–1800 years ago

The present as a warm interglacial

A lucky accident of geological and climatic history has meant that, over the past 8500 years or so (most of the Holocene epoch), our ancestors have enjoyed a relatively stable and warm climate. New data shows however, that the climate was not always as stable as previously thought. There were significant fluctuations that impacted upon humans. Investigation of the most recent geological deposits on land and at sea and the record of ice cores shows that annual average global temperature has mostly been above 15°C (59°F) during the Holocene epoch. This compares with an average of around 11°C (52°F) for the preceding 5000 years. An interesting question which we shall return to is whether the present warm interglacial marks the end of the Quaternary ice ages or whether, despite global warming, it is just another warm phase before glacial conditions set in again.

Ice caps are present at Earth's poles today, but they are relatively small. It was thought that until the very recent phase of melting and retreat which has taken place since the early 1990s, the ice caps and glaciers

had not fluctuated in size since the last major retreat, which began about 12,000 years ago. This time span coincides with the growth of historical records. These records, along with data from ocean sediments and ice cores, reveal important and rapid climate swings which included temporary regrowth of the ice fronts.

Ten thousand or so years ago, modern humans were still hunters who lived in small bands spread around the world. The global human population was probably still only a few million and did not expand greatly until humans changed their way of life to the more sedentary practice of agriculture around 7000 years ago. The postglacial melting of the great polar ice sheets opened new corridors for migration. Animals and humans moved from Asia into the Americas and from the European continent into the British Isles. These land routes were available as long as sea levels remained relatively low; however, as the meltwaters from the ice caps found their way back to the oceans, so sea levels began to rise inexorably. For some populations, there was no return from their offshore islands.

It is no coincidence that many cultures, apart from those of Africa, contain flood legends that record catastrophic natural floods in prehistoric times. The melting of the ice caps released enormous volumes of water, especially into the continental interiors of North America and Asia. Much of this floodwater was unable to drain away, but instead formed vast inland lakes and seas such as the Great Lakes in North America. Landscapes were radically changed, as was their vegetation and wildlife. Remnants of these postglacial environments can still be seen today in the few remaining wildernesses of North America and northern Eurasia. Extensive evergreen coniferous forests, with fragments of cold steppe grasslands, still support large herds of caribou, as they have done since the end of the last Ice Age, about 12,000 years ago.

Some scientists have argued that vegetation and climate change were most likely responsible for the way in which the large populations of medium to large plant-

Hippo skeleton
The discovery of the well-preserved skeleton of a hippopotamus within Ice Age deposits near Cambridge, England showed that at times the climate oscillated widely from cold glacial to warm interglacial conditions. Sea levels also fell sufficiently to reveal land bridges and allow such animals to migrate into the British Isles from continental Europe.

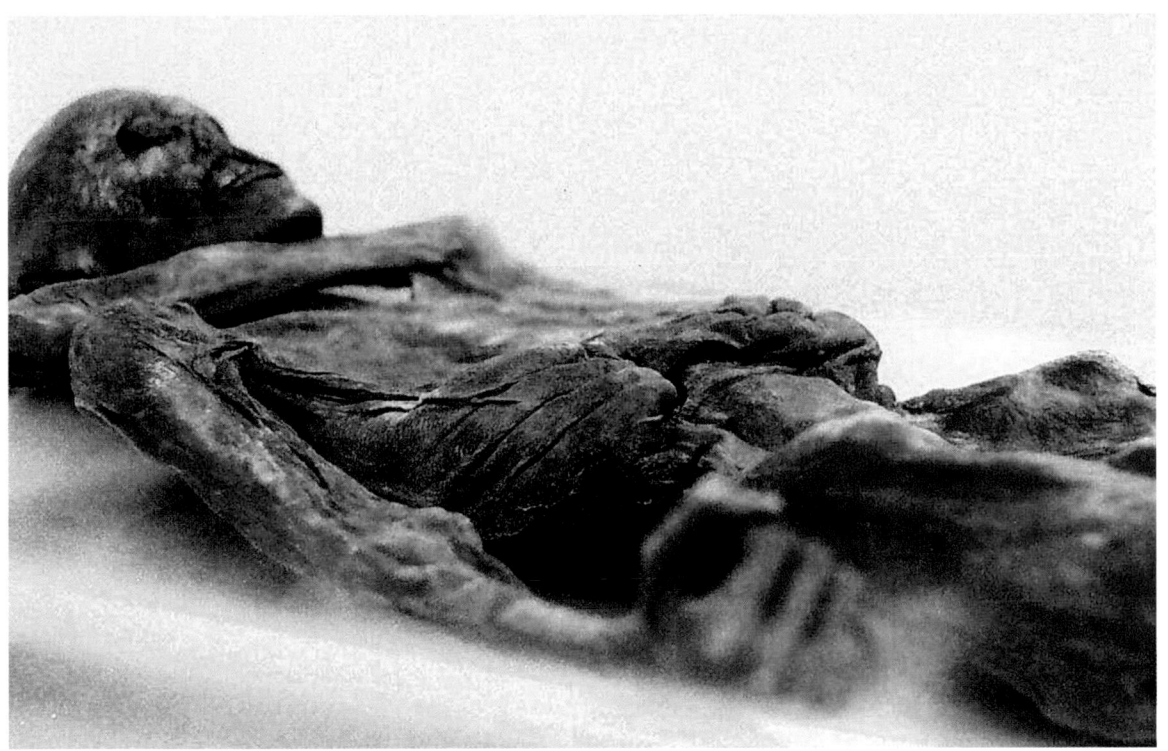

Iceman

The body of a copper-age hunter mummified by glacial ice was found in the Tyrolean Alps on the border between Italy and Austria in 1991. Radiocarbon dating shows that he is 5,200 years old and he is so well preserved that his DNA was recovered and found to be surprisingly close to that of people who still live in the region.

eating mammals crashed all over the world. Certainly many of these herbivores, such as the mammoth, giant deer, woolly rhino, bison, and wild horse, were grazers dependent upon extensive grasslands for their food supplies. Increasing humidity and rainfall soon meant that these grasslands were replaced by forest and woodlands, environments which were more suitable for browsing herbivores, especially deer. But it has to be more than mere happenstance that crashes in populations of these so-called megaherbivores matched the arrival and establishment of growing populations of modern human hunters. And this happened all over the world except in Africa and parts of Asia.

The present view is that it was primarily the arrival of human hunters that caused the megaherbivore populations to crash. With diminishing game available for food, modern humans were increasingly forced to rely more on gathering plant foods and domesticating some wild animals and plants. While the beginnings of agriculture allowed bigger groups of people to live together and the development of settlements, it also required the land to be cleared of forest and woodland. The success of this first agricultural revolution was probably instrumental in promoting population growth

in the more fertile parts of the world. The process still continues today, with little of the postglacial forests, woodlands, and their animal populations remaining. While the rainforests of the tropics may still seem vast, they are only a shadow of their former extent.

Records of recent climate change

It had been thought that the effects of climate amelioration were mostly felt in high latitudes, but it is now known that the tropics experienced major changes as well. Most important was the greening of the Sahara with woodland, lakes, rivers, and abundant wildlife from around 10,000 to 5500 years ago, produced by warmer and wetter conditions. Monsoons circulated rain over the northern tropics and equatorial East Africa. Cooler and drier conditions and the advance of the tropical mountain glaciers followed. These changes are thought to be due to changes in the Earth's orbit around the Sun, which in turn impacted upon atmospheric and oceanic circulation patterns.

Reliable historical climate data is only available for the past 150 years or so, and then only for a few locations, mostly in western Europe. Nevertheless, these data show some interesting and significant trends. For

Cup and ring

These distinctive cup and ring marks were ground into a rock surface by Neolithic people around 5000 years ago in the Orkney Islands of northern Scotland, for reasons as yet unknown to us. They cut across older grooves ground by glacial ice during the Quaternary Ice Age.

the 50 years between 1860 and 1920, the global annual average surface temperature fluctuated by less than 0.5°C (32.9°F). When the average of these 50 years – 15°C (59°F) – is taken as a baseline and compared with more recent years, there is a clear upwards trend which has accelerated since 1980 and now has increased above the baseline by 0.8°C (33.4°F). The global retreat of glaciers verifies this trend. In fact, some may disappear altogether, such as those of Mount Kilimanjaro, which are predicted to melt away by 2015. This trend provides convincing evidence for the existence of global warming at the present.

Records also show that there was a Medieval Warm period from about 1050–1375. In South America, it produced a 300-year drought which may well have destroyed the Tiwanaku civilization around Lake Titicaca, Peru. A well-documented cold phase, known as the Little Ice Age, followed, beginning in the late Middle Ages and lasting for some 300 years until the 19th century. This produced serious crop failures in densely populated regions such as Europe, leading to famine, disease, and civil unrest, culminating in the French Revolution.

Landscape paintings of northern Europe from the 16th and 17th centuries often show frozen rivers and canals with jolly peasants enjoying themselves skating and even holding fairs on the ice. Very rarely are winters cold enough to produce such freezing conditions these

days. These paintings provide striking visual evidence that, over this period, the climate was significantly colder. Such paintings also tend to put a positive gloss on an otherwise socially disastrous period. In upland and marginal areas such as the Scandinavian highlands, Greenland, and Iceland, land was abandoned and populations dwindled. Yet annual average temperatures only declined by less than 2°C (35.6°F).

Wine cultivation, which had been introduced to Britain by the Romans, was abandoned in this region in the middle of the 15th century because of the deteriorating climate; however, it has been renewed since the 1960s. In Europe, data about the wine harvest since the end of the 16th century are remarkably good. Winemaking communities all over France and in some other countries have regularly recorded how many days after 1 September the grapes were ready for harvesting. The warmer and sunnier the growth period, the nearer to 1 September the grapes were ready. This provides a rough proxy measure of climate and shows constant fluctuation with a range of some 20 days every few years. There are also some longer term trends lasting a decade or so.

To gain more accurate details of climate change, however, other biologically based proxy measures can be used. Basically, any organisms which are abundant and sensitive to climate change and can be easily fossilized have the potential for providing such measures. These include trees and their growth rings, which give evidence for climate change over the past 11,000 years; however, well-preserved timber of this kind is rare. Much more common and useful is fossil plant pollen that can be specifically identified and linked to a local flora of a particular time frame (dated by radiocarbon methods which are effective for the past 50,000 years).

The temperature and climate tolerances of living plant species are well known. From these, past climates can be reconstructed as far back as those species have been in existence, which amounts to a few million years for many species. Some animal fossils can be used in the same way, especially certain insects and particularly beetles whose tough carapaces are often preserved in ancient peat and lake deposits of the recent ice ages. More important, however, are certain deep sea-dwelling microorganisms, known as foraminiferans, or forams for short. Analysis of their fossil remains recovered from deep sea sediments has allowed scientists to measure relative changes in the volume of the oceans, which in turn are related to climate and temperature changes. These detailed analyses have been supplemented by data derived from ice cores drilled through the polar ice caps

that have built up over the past 300,000 years or more.

Until about 6000 years ago, human activity impacted mainly upon the biological world. Until this time, tools were derived predominantly from natural organic materials such as wood and bone, and inorganic materials such as flint or obsidian rock materials, from which blades and axe heads were made. The recent discovery of a well-crafted copper axe hafted with yew and dated at around 5000 years old shows, however, that metal extraction from mineral ores must have begun by this time. The axe was found with a frozen body (nicknamed Otzi, the Iceman) in a glacier on the Austrian–Italian border in 1991. Further, the 4000-year-old level within the Greenland ice cores shows a small but distinct rise in dust particles derived from the smelting of copper and other base materials. This demonstrates that metalworking had risen to such an extent that it was already beginning to cause atmospheric pollution. We have escalated this type of pollution to such an extent that there is little doubt that it is contributing to global warming. The stability of the Holocene climate inherited by our ancestors is probably now at an end.

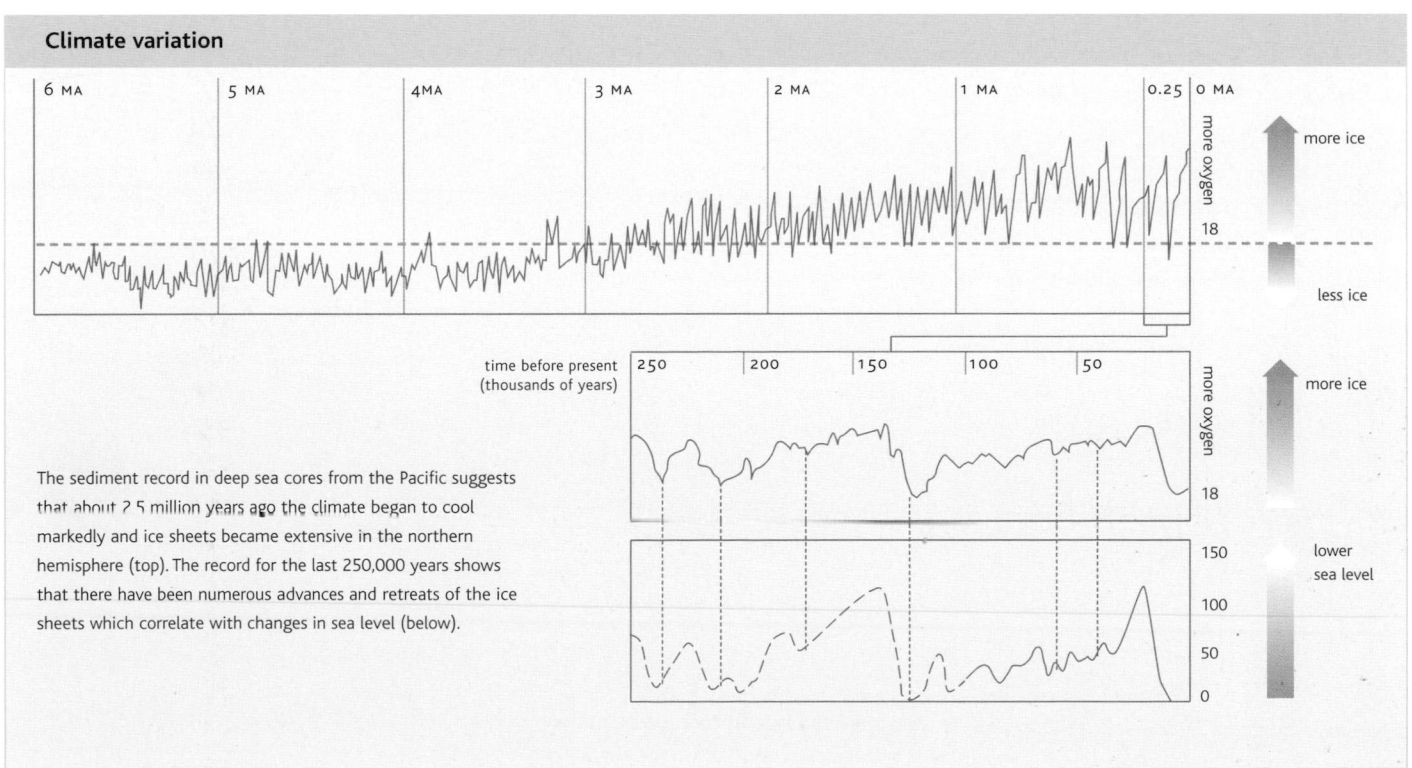

Climate variation

The sediment record in deep sea cores from the Pacific suggests that about 2.5 million years ago the climate began to cool markedly and ice sheets became extensive in the northern hemisphere (top). The record for the last 250,000 years shows that there have been numerous advances and retreats of the ice sheets which correlate with changes in sea level (below).

The discovery of the ice age

Tropical glaciers
At 5895m (19,340ft) high, the snowclad summit of Kilimanjaro, an extinct volcano, looms above the tropical grasslands of Tanzania in Africa. Its elevation is sufficiently high as to maintain several glaciers, although they are now diminishing as a result of global warming.

Today, the existence of the Quaternary ice ages is accepted without question. The idea that over the past two million years there have been repeated advances and retreats of huge ice sheets and glaciers extending out from polar regions into lower latitudes is not a problematic one. The evidence is scattered all over the northern part of North America, northern Europe, and northern Eurasia. Evidence can also be found in the southern hemisphere – the Southern Alps of New Zealand have been sculpted by glaciation into a dramatic landscape of deep U-shaped valleys, and flooded by the sea as fjords, just like the Atlantic coasts of Norway and Alaska.

So how was the existence of the Quaternary ice ages discovered? Since the Middle Ages, large bones found scattered around western Europe in surface deposits were interpreted as belonging to giants. Similar bones had been turned up in classical times throughout southern Europe and the Mediterranean world of Greece, Rome, and even ancient Egypt. Again they were interpreted as belonging to the heroes of classical mythology and often interred in the temples built to honour them.

During the Renaissance, the prevailing world view in the emerging culture of Europe was that of Judeo-Christianity, with a reliance on the biblical texts as historical documents of fact. The power of this world view – which was reinforced by that of the Church authorities and included the power of life and death

Glacial erosion

The effects of glaciation can even be spotted in the tropics when you know what signs to look for. The great (extinct) volcanic edifice of Mount Kilimanjaro – the highest mountain in Africa – near the equator in northern Tanzania still has vestiges of a permanent ice cap and small glaciers. Signs of glacial erosion extend down to 2000m (6500ft) above sea level on its flanks. Today snow and ice are not found below 4000m (13,000ft) on Kilimanjaro, and it is predicted that with global warming the remaining glaciers will disappear altogether in the next few decades.

over heretics – ensured that most naturalists of the day "toed" the party line. Deviation resulted in forced recantations such as that of Galileo Galilei or being burned at the stake like the philosopher Vanini, who dared to suggest that humankind might be related to monkeys. This world view portrayed the development of the Earth and creation of life in very specific terms and added another major historical event, that of the Flood. It was this event that seemed to provide the explanation for the existence of fossils found far inland.

Until the end of the 18th century, most naturalists believed in the literal interpretation of the biblical texts and saw all these fossil remains as evidence for the Flood. Other scholars, however, while still being of a religious persuasion, were becoming more sceptical of such simplistic readings of the texts and of the geological contexts in which the fossils were found. They saw the emerging geological evidence of great thicknesses of fossiliferous strata as proof of numerous flood-type events. By the early 19th century, few had any problems with interpreting fossils as the remains of once-living organisms.

Naturalists gradually came to recognize that some of the gigantic bones found in surface deposits were clearly very similar to those of large living mammals such as elephants. They also realized that, if this were true, they had another problem: how to explain what such animals were doing in northern Europe or North America. One historical explanation for the European finds was that the Romans or Hannibal might have been responsible. It soon became clear, however, that such explanations could not adequately account for the widespread distribution of elephant-related bones nor

their burial in sediment. The discovery of frozen bodies and huge quantities of ivory tusks in Siberia showed that elephant-like animals certainly had lived well beyond the confines of the tropics.

Meanwhile, scientists in European countries bordering the Alps, especially Switzerland, had the opportunity to study glaciers at first hand. Even in the 18th century, scientific exploration of the remarkable phenomenon of glaciers and glaciation was beginning. Erratic boulders often of huge size and quite unrelated to the rock beneath them were seen scattered over the landscapes of northern Europe. Inevitably, the first explanation for their existence and transport was again the Flood. By the early 19th century, however, knowledge of polar glacial phenomena was improving rapidly. It was known that icebergs could carry rocks of almost any size far from their source. British scientists such as Charles Lyell saw sea ice transported by inundations of the sea over the land as the most likely mechanism for carrying such erratics.

Swiss scientists such as Johann von Charpentier (1786–1855), however, began to understand how mountain glaciers were formed. They also recognized glaciers' extraordinary powers of erosion and ability to transport huge quantities of rock debris hundreds of kilometres from mountains down onto lowland plains to be dumped as the glaciers and ice sheets finally melted.

Mammoth
Cave art from the French Grotte de Rouffignac revealed anatomical details of mammoths which were not verified until their frozen cadavers were later recovered from the permafrost of Siberia. Our Palaeolithic ancestors saw and accurately portrayed the length of the hair, the two-"fingered" tip to the trunk, the hump of fat on the shoulders and head, and the distinctive sloping back of these extinct elephant relatives.

Zermatt glacier
A 19th century engraving illustrates the growing interest in glaciers, their structures and mechanisms of erosion. The trains of medial and lateral moraine rock debris are clearly shown on the surface, while a falling boulder depicts how further debris is accumulated through rock falls from the precipitous flanking rock walls.

movement. Glacial erratics – boulders carried far from their original sources by the ice and dumped wherever the ice finally melted – were described and traced back to source. Eventually some very distinctive and unique rock types were found which could be used as reliable indicators. For instance, boulders of an unusual kind of granite called ailsacraigite after the island of Ailsa Craig (technically a riebeckite microgranite) off the western coast of Scotland were found to have been carried as far south as Dublin on the east coast of Ireland. Scandinavian metamorphic rocks were found scattered over the North German plain and down to the northeast coast of England.

Gradually all the classic erosive features of upland glaciation and the largely depositional ones of lowland glaciation were revealed – the cirques (also known as corries, or cwms); the U-shaped glacial valleys with their "hanging" tributaries; the roches moutonnées; and the depositional features such as lateral and terminal moraine, drumlins, kettle holes, eskers, outwash sands, and gravels. Agassiz emigrated to America in 1846 and carried on his great glacial conversion campaign. All the same features were to be found around the Great Lakes and everywhere to the north, and on a much grander scale than seen in Britain and even continental Europe. Evidently, high latitudes of the northern hemisphere had suffered a major climate change in the not-too-distant geological past. The question was, why?

The idea that major climate change such as an ice age might be connected to the elliptical orbit of the Earth around the Sun was first promoted by a French mathematician, Joseph Adhémar (1797–1862); however, it was a remarkable Scot James Croll (1821–90), who really developed this concept scientifically.

As geologists investigated the land-based record of glaciation, they realized that there was definite evidence for alternations of glacial and warmer interglacials. For instance, the discovery of a hippopotamus skeleton buried in glacial deposits by the banks of the River Cam in Cambridge, England, provided striking evidence that some interglacials must have been even warmer than present climates at these latitudes. Equally, the discovery of fossils of woolly mammoths in the same area of England showed just how drastically the climate had changed. The problems lay in determining how

One man who became converted to the glacial ideas of Charpentier and others was the Swiss palaeontologist and expert on fossil fish, Louis Agassiz (1807–73).

Armed with this evidence of glacial processes and mechanisms, Agassiz showed British geologists such as William Buckland (1784–1856) and Charles Lyell how to spot the signs of past glaciation in the landscape. Soon the scientific literature was filled with descriptions of ice-scratched and grooved rock surfaces, with the alignment of the marks indicating the trend of ice

James Croll (1821–90)

James Croll was a brilliant and largely self-taught scientist. He worked as a millwright, carpenter, shopkeeper, and insurance salesman before settling in Glasgow, aged 36, as a janitor in the Andersonian College, a job which gave him access to an excellent library and time to read as much science as he wanted.

As a result of these studies, Croll combined measurements of the ellipticity of the Earth's orbit and regular changes of the equinoxes to calculate how the poles could regularly suffer cooling sufficient to generate periodical ice ages, with warmer interglacial periods in between. He also discussed the role and impact of ocean currents on climate.

His ideas, first published in 1864, had a big impact among the scientific world, and soon elevated Croll into a professorship and fellowship of the Royal Society, one of the oldest surviving scientific societies in the world, which had been founded in London in 1660.

many climate swings there had been and over what interval of time.

The task of trying to match sequences of fossiliferous interglacial deposits, mainly lakebed sediments, between different regions of northern Eurasia was difficult enough, let alone trying to match them with North American sequences. The problems of dating, even when radioisotope dating was available, were also enormous. Each glacial phase had tended to wipe the slate clean by removing many previous glacial-related deposits. Most of those that survive belong to the last glacial phase. Persistent searching over the decades, however, has managed to uncover some older ones, and correlation has been greatly helped by the discovery that the best records of climate change during the Quaternary ice ages actually lie at the bottom of the oceans and locked up in the polar ice sheets.

Glacial flow

Minor glaciers, flowing down from the sites of origin in high bowl-shaped cirques, coalesce with main valley glaciers like tributary rivers. Lateral moraines from the side glaciers are amalgamated to form medial moraine in the main flow. Evidence showed that the glaciers had been much more extensive in the past.

Evidence of life during the ice age

More than 200 hundred years of finding fossils in superficial deposits on the land have revealed astonishing insights into the life of the Quaternary ice ages. Many of the giant bones scattered over Eurasia and North America were, as we have seen, originally interpreted as belonging to giants of myth and folklore. They were then seen as victims of the biblical Flood, or Deluge, before finally being recognized as what they really are: representatives of the flora and fauna that lived and died as the climate swung back and forth between cold glacials and warm interglacials. By the time the ancient Egyptians were building the early pyramids some 4800 years ago, however, a significant proportion of the Ice Age animals – including at least one human species, *Homo neanderthalensis* – had become extinct. For this was a critical period in the evolution of modern humans. Over many years now, scientists have been arguing about the cause of these extinctions. Was it climate change, the hand of modern man, or a combination of the two?

It is still astonishing to realize just what the landscapes of the Quaternary must have looked like to our human ancestors and especially the first modern humans who were the first humans to enter the Americas (some 15,000 years ago) and Australasia (some 60,000 years ago). The great continents of Eurasia, the Americas, and Australasia were like the best of today's African game reserves, with a wonderful diversity of animals and plants. By 10,000 years ago, however, they were severely depleted of large game; by 4800 years ago, when the last of the mammoths died out, there were very few large mammals left. Seemingly blind to what they were doing, humans continued to decimate the survivors until today's impoverished fauna was "achieved". Curiously, in much of Africa and parts of Asia, modern humans have, until now, managed to coexist with wild and often dangerous animals.

The evidence for the diversity of life during the Quaternary ice ages is various and ranges from fossil bones to ancient biomolecules. Much of the story is indeed told by fossil bones. For instance, since the early 1970s, the skeletal remains of some 50 Columbian mammoths (*Mammuthus columbi*) have been excavated at Hot Springs in South Dakota. Dated at around 26,000 years old, the bones were found jumbled together in sediment at the bottom of a deep water-filled natural hole (called a sinkhole) formed in the local limestone. The water attracted thirsty beasts, some of whom fell in and could not get out because of the steep sides and eventually drowned. The Columbian mammoth first entered North America via the Bering land bridge one and a half million years ago and was not as adapted to the cold as its relative the woolly mammoth (*Mammuthus primigenius*). Interestingly, the remains of a single woolly mammoth have been found at Hot Springs, so the two may have occasionally encountered one another.

A somewhat similar deposit at Condover in Shropshire, England, was found in 1986. From a lens of clay within a gravel pit, some 400 bones were recovered and dated at between 12,700 and 12,300 years old towards the end of the last glacial. When the "jigsaw" was put together, there was the best part of the skeleton of an adult male mammoth, aged around 28 years, and three babies between three and six years old. Again they were probably attracted by water and plants to a treacherous waterhole and drowned. Here it was a steep-sided kettle hole formed by the melting of a buried mass of ice which subsequently filled with water. Insect remains recovered from within the bones show that, after death, the bodies floated, and flies laid their eggs on the rotting corpses. Other beetle and plant-pollen fossils were used to match the muddy sediments to other similar deposits which developed as the ice gradually retreated into the mountains of nearby Wales.

The large bones of large land mammals such as the mammoth may be spectacular, but they are far from being the only evidence. Thanks to the remarkable deepfreeze condition of the polar permafrost, whole cadavers are still being recovered of animals which lived as long as 70,000 years ago. The mammoths may be the most famous of the permafrost cadavers, but there have also been woolly rhinos, bison, horses, and wolverines recovered from the permanently frozen ground of Siberia and Alaska.

In 1979, an Alaskan gold miner sluicing placer gold from permafrost near Fairbanks uncovered the legs of a large animal. Fortunately he alerted scientists, who were able carefully to excavate the 35,000-year-old remains of a bison which became known as Blue Babe. Deep scratch marks on his rear quarters and puncture wounds showed that he was killed at aged eight or nine years by a very large cat, bigger than the surviving cougar who

used the same killing technique still employed by lions. The cat jumped onto the bison's back, holding on with its claws, and suffocated its victim either with a muzzle or throat bite.

If palaeontologists had to rely just on the bones of Ice Age animals, they would only have been able to guess at how these animals were adapted to living in cold, subpolar climates. The frozen cadavers provide direct evidence of just how well they were adapted for the conditions. Like the living musk ox, which can tolerate subzero conditions for a certain period, the woolly mammoth and woolly rhino, as their common names suggest, not only had skin tissues thickened with fat deposits, but also a thick layer of insulating hair. There was a woolly underhair covered with much longer, straighter outer guard hair, the individual hairs of which were thicker and tougher than the underhair. The mammoth also had much smaller ears and tail than

Warmth in the Ice Age

Fossils found within Ice Age deposits, such as those found during building work in London's Trafalgar Square, reveal that, at times during the Ice Age, life was similar to that of today's African game parks. Straight-tusked elephant, hippos, large bovids, deer, bear, and big cats all occupied the open wooded landscapes.

modern elephants. These particular features have been independently verified thanks to the portraits of mammoths provided by our ancestors some 30,000 years ago.

The frozen cadavers of the permafrost are often so well preserved that the soft tissue, gut contents, hair, and DNA can be recovered. Despite recent hopes that mammoths might be resurrected by cloning processes using ancient DNA and living elephant mother surrogates, the chances of viable offspring being produced are negligible, and in any case they would be chimeras, mostly elephant and only partly mammoth. Nevertheless, the recovery of such ancient DNA from cold climates is in itself very interesting and shows that

the modern elephant and mammoth evolutionary lineages diverged from one another some five million years ago. The recovery and analysis of ancient DNA are proving invaluable for resolving other problems as well.

It was not just the famous and familiar beasts of high latitudes that became extinct towards the end of the Quaternary ice ages. The lower latitude faunas of the Americas and Australasia were equally as badly hit. Unique localities such as the tar pits of Rancho La Brea, in today's downtown Los Angeles, California, have provided wonderful insights into life well beyond the ice between 40,000 and 10,000 years ago. Excavation of this site over many decades has uncovered some one and a half million bones, weighing 100 tonnes (98 tons),

Baby mammoth
The remarkably well preserved body of an emaciated 18 month old baby mammoth was found in the frozen permafrost ground of a Siberian gold mine in 1977. Nicknamed Dima, it was found to be 40,000 years old by radiocarbon dating.

Frozen Mammoth

In 1901 an expedition from St Petersburg recovered as much as it could of the frozen remains of a large bull mammoth which died around 30,000 years ago when it was about 35 years old. The entire skeleton, much of the skin, tissue, and hair, apart from that of the head, which had been scavenged by foxes, was recovered and reconstructed in St Petersburg Zoological Museum, where it can be seen today.

and two and a half million invertebrate fossils. The animals range from giant ground sloths and sabre-toothed cats to beetles. La Brea was another natural trap like the sinkhole found at Hot Springs, this time filled with sticky tar. The herbivorous animals were attracted by plants, mired in the tar, and eventually died. Their carcasses attracted predators of all sizes from big cats to vultures, and many of these also became trapped. What was bad luck for them has been good fortune for scientists, who have been able to reconstruct a detailed picture of the life of the region. One of the interesting aspects was the inclusion of animals from South America.

Life in the southern hemisphere was just as interesting, with some unique faunas and floras occupying Australia and the innumerable offshore islands of the Pacific. New Zealand is particularly intriguing because it separated from Australia at least 80 million years ago and "drifted" into the Pacific with a "cargo" of plant and animal inhabitants that continued to live in a time warp. New Zealand's South Island was severely glaciated because of its high latitude and the development of a range of high-altitude alpine mountains. The cold climate eliminated many of the eucalyptus trees, but left numerous ancient kinds of conifers called podocarps and araucarias which have changed little over the past 190 million years. One of the latter is the kauri, which holds the world record for volume of timber in a single tree – the largest recorded kauri had a girth of 23.43m (76.87ft).

In New Zealand, there has been no significant climate change over the past 1000 years. Its unique animals include ancient kinds of frogs, the tuatara lizard, giant earthworms, crickets, and carnivorous snails, along with an amazing diversity of birds. Some 245 bird species, mostly unique, occupied the islands until the first arrival of humans only 1000 years ago. They included many flightless species and some giants such as the moa (*Dinornis novaezealandiae*) which weighed around 98kg (216lb). Some 40 of those species are now extinct thanks to the human introduction of rats, hunting, and agriculture.

Human prehistory

Today, we know from the fossil record that there have been at least 15 and perhaps as many as 20 extinct human relatives. To date, the oldest known human relative is the seven-million-year-old *Sahelanthropus tchadensis* found recently in the central African state of Chad. Furthermore, our human family "tree" is bound to increase as more bones and stones are uncovered. No longer can we see our human ancestry as a single line of evolution from an ape-like ancestor walking on all fours and becoming progressively more human-like with an upright stance. The picture is much more interesting and complex, with the family tree acquiring a bushlike appearance, the evolutionary interconnections of which are unclear. Our ancient relatives appear as fossils with mosaics of more and less advanced characters. As often as not, they disappear without apparent issue. Just how much of this fragmentary jigsaw is an artefact of poor preservation by the fossil record has yet to be established.

But bones and stones are not the only line of scientific evidence for our prehistory. Advances in genetics and molecular biology have opened up whole new lines of investigation and data. Mapping of the human genome verifies our close evolutionary connection to the chimpanzee and the recentness of a shared ancestor who lived in Africa a mere seven million years or so ago. Future mapping of the chimp genome, now under way, promises to show how some of the critical characters that separate us from the chimpanzees have evolved and what genes were involved in these evolutionary changes. Most importantly, comparison of the DNA from human populations around the world shows differences that are so slight that all modern humans must have a common origin in Africa as recently as 100,000 years ago.

The recovery of ancient DNA from extinct fossil relatives, the Neanderthal people of Eurasia, verifies the genetic separatedness of our species, *Homo sapiens*, from the Neanderthals. And yet the Neanderthals were also very successful "people" who survived the often harsh conditions of the Quaternary ice ages for more than 300,000 years. They had culture and speech, although probably not complex language, and yet they died out around 30,000 years ago. Their similarity to modern humans questions the whole notion of when exactly "humanness" evolved and what it was that gave *Homo sapiens* that extra edge which led to our remarkable reproductive success as a species. Two hundred years ago, virtually none of this information was available, although there were suspicions in the minds of some naturalists that humans might be related to the apes. They also thought that humans might have a prehistory very different from that portrayed in biblical and other religious texts with their various Creation stories.

From the early days of exploration and discovery of the diversity of life on Earth, it became clear to the predominantly European explorers that the animals and plants found in Eurasia were far from being typical of the rest of the world. Tales of strange ape-men were given reality when, in the 17th century, live apes (orang-utans) were brought back to Europe from Southeast Asia by Dutch explorers and colonists. Not long after, chimpanzees were brought back from West Africa, and, by 1698, a London physician, Edward Tyson (1651–1708), had the opportunity to dissect one such unfortunate captive chimpanzee, which had died soon after its arrival in England. Tyson showed how similar the anatomy and skeleton of the chimpanzee is to that of humans. Indeed, he listed some 50 common characteristics.

By the 18th century, the Swedish pioneer taxonomist Carolus Linnaeus (1707–78) grouped apes and humans together in the order Anthropomorpha, later changed to order Primates, where we and the apes still belong. Although criticized, Linnaeus challenged anyone to find significant anatomical differences between the two that would justify separating them. It is important to realize that Linnaeus was a firm believer in the Christian doctrines and considered that, in his work, he was merely illustrating the grand design of God's work and Creation. He firmly believed in the fixity of species.

By the end of the 18th century and the beginnings of the 19th century phase of the scientific revolution, the pace of discovery increased. In 1801, John Frere (1740–1807), an antiquarian from Norfolk in England, published the first illustration and description of a flint hand axe discovered in surface deposits near his home. He recognized that the shape of the axe could not be accidental, but was a clear sign of the hand of man and that humans must have a much more ancient history than previously recognized. It took another 50 years and the discovery of many more stone tools, especially in France, before there was any acceptance by scientists that humans had lived alongside the extinct animals of the Ice Age. Even the discovery of the first fossil remains of the extinct Neanderthal people in 1856 was not accepted until the 1880s.

By the 1870s, while some scientists had begun to accept our evolutionary connection to other animals and especially the apes, there was widespread argument about where our ancestry lay because there were no fossils. One of Darwin's most energetic supporters was the German biologist Ernst Haeckel (1834–1919), who did not always agree with his hero. Haeckel argued that humans were more closely connected to the apes of Southeast Asia, where the "missing link", the fossil evidence connecting the apes and humankind, would be found.

Eugene Dubois (1858–1940), a young Dutch anatomist, was so enthralled by Haeckel's argument that he determined to be the person to prove him right. In the late 1880s, he gave up his university post, uprooted his family, and set out for Java, having enrolled as a military doctor in the Dutch colonial forces there. After several years searching, he uncovered some bones which he thought proved Haeckel correct. Dubois claimed that the missing link was an upright, walking apelike man with a relatively small brain named *Pithecanthropus erectus* (now regarded as *Homo erectus*), who lived alongside the extinct animals of the Quaternary period. Again the scientific community was reluctant to accept this startling new evidence for another extinct human relative, especially one that lived in Asia.

We now know that Dubois was partly right. He had indeed found a very remarkable ancient relative, but one

T.R. Underwood, del. 1797.   *Flint Weapon*

that did not evolve in Asia. Our *Homo erectus* relatives evolved in Africa, as Darwin had suspected, in early Quaternary times. What is remarkable is that some of them moved north out of Africa and managed to spread throughout much of Asia within a few thousand years. Theirs was the first great human diaspora, and they achieved it armed only with a basic tool kit of stone axes and simple stone blades. They thrived in eastern Asia and may have survived until as recently as 50,000 years ago.

First axe

This beautifully fashioned flint handaxe was the first ever to be ascribed as the work of the human hand. Found in the east of England, it was illustrated and described in 1800 by the antiquarian John Frere. It was probably the work of Neanderthals or early Cro-Magnon people.

Back in Africa, *Homo erectus* gave rise to further populations of migrants which moved into Europe and parts of Asia. Generally recognized as species such as *Homo antecessor*, *Homo heidelbergensis*, and *Homo neanderthalensis*, these relatives had bigger brains and made somewhat more advanced stone tools. But again it was back in Africa around 200,000 years ago that another population arose with more modern human biological characteristics and developed the greater cultural skills for which they are recognized as members of our species *Homo sapiens*. They spread throughout Africa and eventually splintered into a number of populations whose descendants still can be recognized in Africa. Recent discoveries from southern Africa show that, by 90,000 years ago, these people were using sophisticated tools made from bone, were burying their dead with ceremony, and had begun to use graphic symbols.

By this time, some of them had also made their way north out of Africa to the Middle East, where they encountered the incumbent Neanderthal people. Around 60,000 years ago, our African ancestors spread throughout central Asia and got as far as Australia by around 50,000 years ago, perhaps encountering some surviving *Homo erectus* relatives on the way. In western Europe, modern humans and Neanderthals came into close proximity and overlapped for about 10,000 years. It was thought that the Neanderthals may well have been the ancestors of the original Europeans, but ancient DNA evidence discounts this. It has been discovered that the Neanderthals possessed a few genetic signatures which distinguish them as a separate interbreeding species. None of these genetic signatures turns up in the modern European gene pool.

The extinction of the Neanderthals is just like that of the other large Quaternary mammals closely coincident on the arrival of modern humans. That is not to say that there was blood on the ground from widespread conflict. It is more likely that the modern humans outhunted the smaller bands of Neanderthals, whose population gradually dropped below sustainable numbers, just as is happening today to many of the remaining large mammals, such as the rhinoceros and tiger, as a result of hunting, poaching, or the destruction of their habitats by modern man.

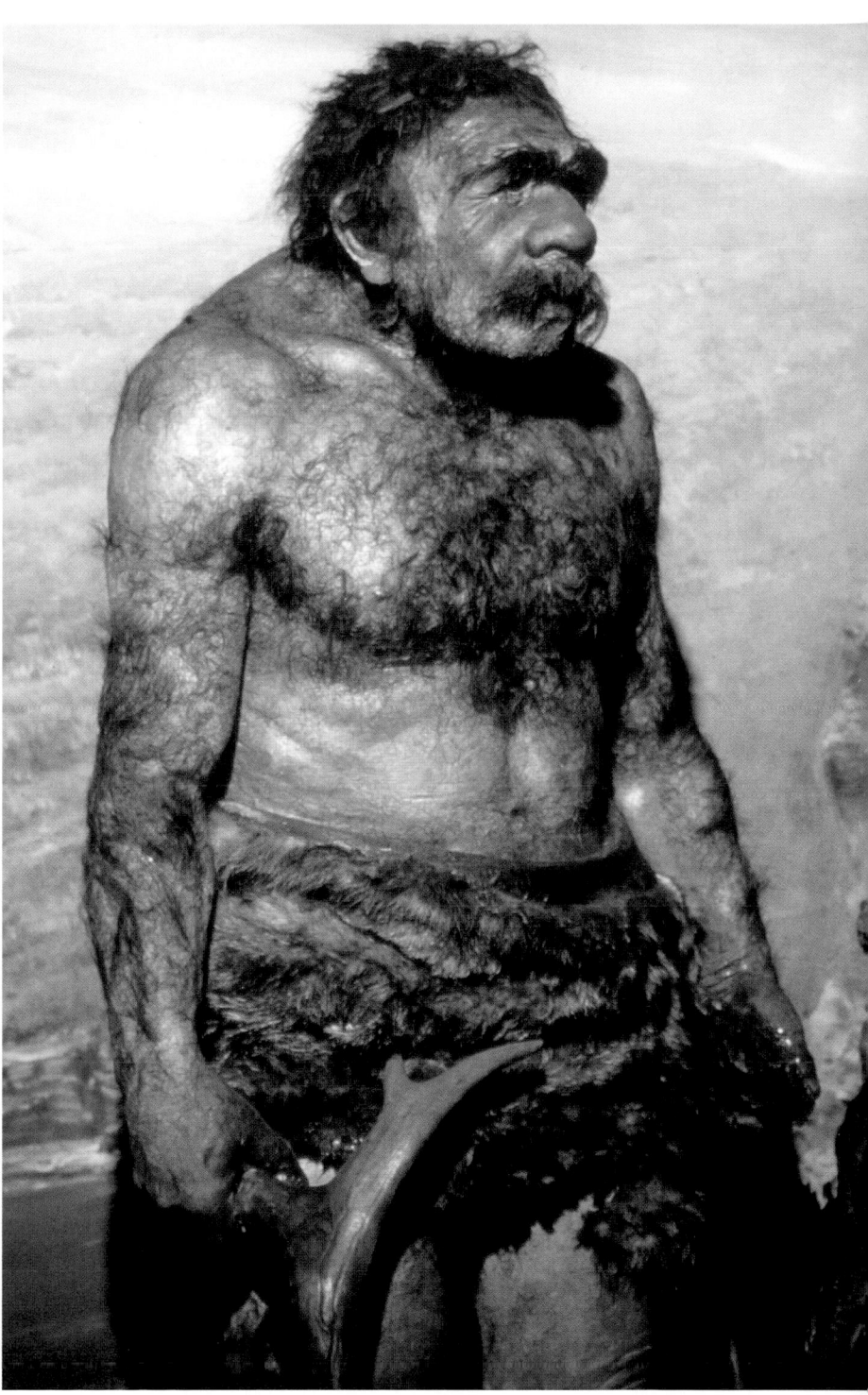

Neanderthal talking point (left)
The most complete skeleton of a Neanderthal is that of an adult male, 60,000 years old, found at Kebara, Israel, which preserves evidence that they had the power of speech.

Neanderthal man (above)
Powerfully built and big brained.

The environmental impact of the ice age

The "toings and froings" of the glaciers and ice sheets of the Quaternary Ice Ages had a devastating effect upon the polar realms of North America and Asia, Antarctica, New Zealand, Tasmania, and southernmost tip of South America. Remnants of these ice bodies are still to be found at the poles and at high altitudes in lower latitudes such as in the Himalayas, Andes, and the tropical world. During glacial periods, both the polar ice sheets and mountain glaciers spread over large areas and had a significant effect on local and regional climates. The erosive effects of glaciation on upland landscapes are well known. Less well known is the effect of glacial deposition and reworking by glacial meltwaters. Even less well known is the impact of the ice ages on the tropics.

Beyond the furthest reaches of the ice, there was a vast swathe of permafrost where the ground was frozen all year apart from a thin surface layer which melted during summer. The permafrost extended for hundreds of kilometres beyond the ice and had a major impact upon vegetation and life. For instance, most of the southern part of the British Isles, which lay beyond the ice front at its maximum, was within the permafrost zone. Developing river valleys cut into limestone terrains that normally do not support surface water. These old drainage patterns are still visible today, even though with the melting of the permafrost the surface water has retreated underground once more to leave dry valleys.

Beyond the permafrost, all the vegetation zones were displaced towards the equator during the coldest glacial spells and retreated towards the poles during the warmer interglacial periods. Consequently, there was a constant shifting of plants with which the animals had to try to keep up. For instance, in the Mediterranean region during cold glacial periods, hardy conifers more typical of Scandinavia replaced the plants such as native vines and citrus. The life that such hardy plants support is very different from the warm Mediterranean fauna that includes several kinds of animals intolerant of cold, especially deer, wild cattle, sheep, goats, and reptiles such as tortoises, lizards, and snakes.

The continental interior of Eurasia and North America also had vast cold deserts with little or no snow or rain, and consequently no vegetation. Soils and rock were exposed to freezing temperatures and winds which dehydrated the surface and blew away any small-sized loose rock particles. This rock and soil dust was carried by the prevailing winds until their speed and energy fell and the dust fell back to Earth. So much accumulated in some places, especially in parts of China, that it blanketed the landscape with many metres of fine silt and clay deposits known as loess.

Altogether some 160 significant climate shifts (80 glacials and 80 interglacials) over the past two million years of the Quaternary ice ages have been documented from deep-sea sediment records. The on-land records of most of these have been removed by subsequent erosion, but the net effect has been to erode high ground and dump much of the eroded rock material on lower ground as the ice melted. Renewed advances over lower ground produced ice sheets and permafrost conditions which reworked and patterned existing deposits.

With the final "meltdown" beginning some 12,000 years ago, meltwaters produced enormously powerful

Retreating glacier
As glaciers melt as a result of global warming, they reveal the impact they have had on the landscape. Here moraine rock debris lines the valley sides but downslope, lichens, mosses, and tough grasses begin to colonize the debris and build up soils which will allow the future growth of large shrubs and eventually trees.

rivers and floods which again reworked much of the glacial deposits into vast outwash plains of sands and gravels, strewn with boulders and clay-filled hollows. Enormous tracts of Eurasia from the Baltic plains eastwards towards the Pacific were affected. Equally, in North America, huge areas south of the Great Lakes were also affected, with flood events forming extraordinary landscapes such as the Channelled Scablands of Washington and Oregon. To the east in Montana lay the vast ice-dammed Lake Missoula, measuring some 7700km² (2,970 square miles) and 300m (1,000ft) deep. When the ice melted, some 20,000,000m³ (720 million cubic feet) of glacial lake water flowed out in a couple of days and travelled westwards as far as Portland, Oregon, and the sea. The flood cut channels down to bedrock in places and redistributed glacial debris in huge waves of sand and gravel. This process happened many times as the lake

refilled and flooded out. No doubt much life was destroyed, perhaps including some of the early human inhabitants of the Americas.

The effects of climate-driven shifts were also felt much closer to the equator and impacted upon our human ancestors and the animals they hunted. Recovery of ice cores from tropical mountain glaciers such as Peru's 6048m (19,843ft) Huascarán dating back to 19,000 years ago and Bolivia's 6542m (21,463ft) Sajama reaching back to 25,000 years ago is telling a remarkable story. The ice from here shows that South America was between 5° and 12°C (41° and 53.6°F) cooler then than it is now, and the Amazon basin was probably deforested. Andean lake sediments tell the same story. Indeed, some scientists now claim that it is the tropics that drive climate change because so much of the Earth's landmass and thus its water recycling at present lie in the tropics.

Medial Moraine

A close-up of fresh rock debris forming a medial moraine shows how large angular boulders can be carried considerable distances from their source by the conveyor-belt transport mechanism of glaciers. As the debris is removed, more fresh rocks are exposed to weathering and erosion.

3
Life's third age

Apart from scattered Quaternary age surface deposits which provide the introduction to Earth history, the first layers of this geological story, recount the events of life's third age – the Tertiary as it was originally called. We now divide its history into two periods – the Neogene (1.8–23.8 million years ago) and the Paleogene (23.8–65 million years ago). Together, they are sometimes referred to as the Age of Mammals, although it was just as much the age of flies, flowers, songbirds, modern bony fish and primates. As we shall see, life's third age began abruptly with a bang.

Because these young "Tertiary" deposits generally lie close to the surface, especially in the Mediterranean region, they play an important role in the history of geology. Many of these rocks and their contained fossils were uplifted from the surrounding seas by earth movements associated with the building of the European Alps. This accident of geological history was a lucky one because it meant that many of their fossils were not greatly different from life forms found today.

	3800 MA The first evidence of chemical life on the planet	
PRECAMBRIAN EON		
ERA	4600 MA HADEAN	3800 MA ARCHEAN

CENOZOIC ERA		
PERIOD	65 MA PALEOGENE	
65 MA End Cretaceous extinction	56 MA Diversification and radiation of mammals	41 MA *Eosimias*, the earliest anthropoid
		40 MA Opening of North Atlantic, and volcanism
		38 MA Break-up of Australia and Antarctica
		35 MA Impacts at Chesapeake, USA, and Popagai, Russia

Among these Tertiary fossils, the remains of shellfish, clams, snails, and sea-urchins, as well as sharks' teeth were all clearly recognisable when they were discovered by geologists. The similarities to modern life forms helped these scientists to recognize the fossils as the remains of past life.

The Mediterranean region is still subject to reminders of Earth's dynamism. Earthquakes and volcanic eruptions are common and can easily be linked to particular rock products such as lavas and geological phenomena such as faults. It took centuries for such phenomena to be traced to fundamental driving mechanisms within the earth. Geological wonders such as the Giant's Causeway and Fingal's Cave were celebrated by artists (for example the German composer Felix Mendelssohn, 1809–47) for hundreds of years, but only recently has the Paleogene outpouring of these lavas been linked to the opening of the North Atlantic Ocean, which occurred as the ocean floor spread through the dynamic of plate tectonics.

Fossil stag beetle

Stag beetles aré very rare as fossils, let alone specimens as well preserved as this one which still retains some iridescence on its carapace. Comparable forms are today only found in the tropics of Southeast Asia but this one was found at the 49 milion year-old Messel World Heritage Site in Germany.

1200 MA The first multi-celled organisms date from the middle of the Proterozoic period

610 MA The first large marine animals appear

PHANEROZOIC EON

2500 MA PROTEROZOIC | 545 MA PALEOZOIC | 248 MA MESOZOIC | 65 MA CENOZOIC

TODAY

23.8 MA NEOGENE | 1.8 MA QUATERNARY

20 MA India collides with southern Asia; building of Himalayas

11 MA Late Miocene climate cooling; spread of grasslands

6–8 MA Common ancestor of chimps and humans

3.6 MA Laetoli bipedal footprints

3–3.8 MA *Australopithecus afarensis* (Lucy)

2.4–3 MA *Australopithecus africanus*

2.5 MA Panama freeway and fauna exchange

2.65 MA Oldest stone tools

Discovering the ancient testimony of rocks

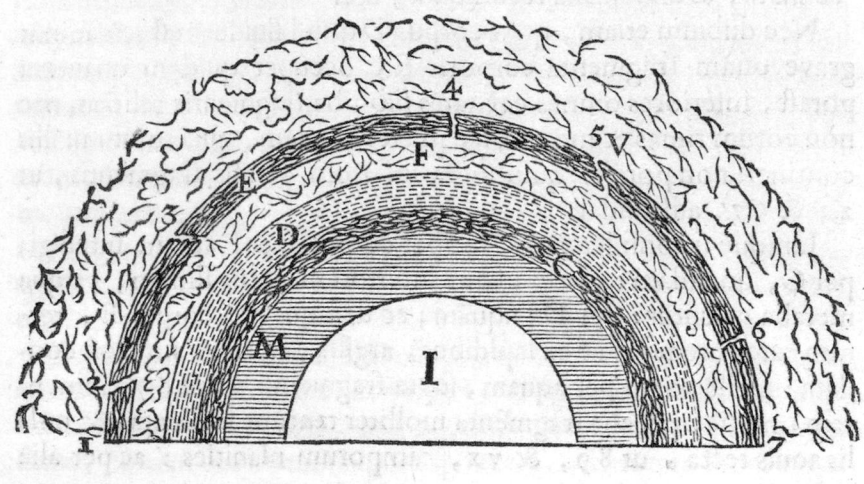

PARS QUARTA. 155

semper augerentur, tandem ejus partes tam parum fibi mutuò ad-
hæferunt, ut non ampliùs in modum fornicis inter F & B poffet fuf-
tineri, & ideò totum confractum, in fuperficiem corporis C gra-
vitate fuâ delapfum eft. Cumque hæc fuperficies fatis lata non effet,
ad omnia illius fragmenta fibi mutuò adjacentia, & fitum quem prius
habuerant fervantia, recipienda, quædam ex ipfis in latus inclinari

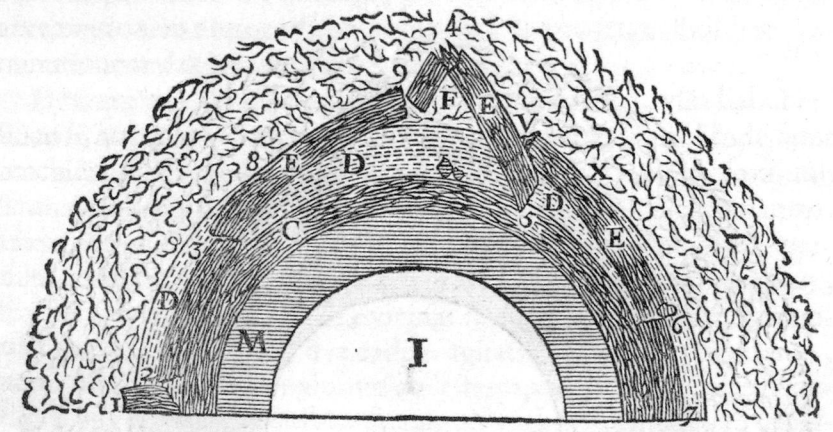

atque una in alia recumbere debuerunt. Nempe fi exempli gratia,
in eo tractu corporis E, quem hæc figura repræfentat, præcipuæ
fiffuræ ita fuerint difpofitæ in locis 1 2 3 4 5 6 7, ut duo fragmen-
V 2   ta

Following the early discovery of fossils in rock strata, the nature of fossils and where exactly they occurred became a particular matter of debate in Europe from the 16th century onwards. By this time, natural historians had also recovered some of the texts of earlier scholars of the classical era, who had themselves questioned the nature of fossils. Much of the 16th-century debate concerned the exact identity of the creatures to which the ancient scholars had been referring, because none of their texts was illustrated. By the latter part of the 17th century, the debate had moved on, and fossils were seen in the somewhat wider context of the rocks in which they occurred. Big questions were asked about how the Earth had formed, how rocks had been crushed to produce mountain ranges, and what fossils were doing so far inland when clearly many of them appeared to be the remains of sea-dwelling creatures.

One of the favourite explanations was that all rocks were deposits of some primal ocean – this seemed the best explanation for the occurrence of so many rocks in layers. Accordingly, those at the top of the pile would be relatively younger than those at the bottom, where crystalline granites and other ore-bearing rocks were thought to have formed first. By the mid-18th century, German experts on minerals and mining such as Johann Gottlob Lehmann (1719–67), Georg Christian Füschel (1722–73), and Abraham Werner (1750–1817) suggested that there was a threefold division of rocks: first, those that had been elevated or folded to form high mountains and often included ore-bearing rocks; secondly, those that formed during the Noachian Deluge, contained fossils, and were originally flat-lying; and finally those that were post-Deluge, also flat-lying, but younger. An Italian contemporary, Giovanni Arduino (1714–95), named these divisions as Primitive, Secondary, and Tertiary.

Primitive rocks were non-fossiliferous ones such as granites, basalts (commonly forming lavas), schists, and gneisses (altered by heat and pressure to form metamorphic rocks), all of which were characteristically

PERIOD | PALEOGENE | NEOGENE | QUATERNARY

EPOCH | 65 MA PALEOCENE | 34.8 MA EOCENE | 33.7 MA OLIGOCENE | 23.8 MA MIOCENE | 5.3 MA PLIOCENE | 1.8 MA PLEISTOCENE | 10 KA HOLOCENE

found in high mountains such as the Alps and contained valuable ore minerals. The Secondary rocks were stratified sandstones, shales (compressed mudstone), and limestones, which were often highly fossiliferous. They tended to be found in the flanks of high mountains and formed lower hills such as the Jura Mountains. Finally, the Tertiary rocks were also stratified and formed a younger series of sandstones, mudstones, and limestones which were also highly fossiliferous. They could also form low hills and were sometimes made of sediment derived from the underlying Secondary rocks.

It was these Tertiary rocks which were particularly common in northern Italy where Arduino worked. The form of rock strata in the same region had, in the previous century, inspired the Danish scholar Niels Stensius (more generally known as Nicolaus Steno) to consider how valleys and hills were formed. He was one of the first to clarify the relative ages of successive rock strata. He also suggested that strata could be deformed either by volcanic activity or by their collapse into cavities below ground. The idea that only gravity was powerful enough to explain the folding and faulting of rocks was not unreasonable for the time. There were no known forces that could push mountains upwards, but there was plenty of evidence to show that gravitational collapse could be immensely powerful and catastrophic.

With the beginning of the Industrial Revolution in the 18th century, exploration of naturally occurring materials with potential economic worth reached fever pitch in Western Europe and was gradually spreading elsewhere. There was plenty of easily workable rock available at the surface for building and road making. Common rock materials such as limestone had additional roles in lime making and even lithography, while clays and mudstones were used for pottery and brick making. Other valuable rock materials such as coal and iron ore, however, were found in only a very limited number of surface sites.

The realization that seams of stratified rocks could be pursued underground from surface outcrops was a major breakthrough. It was also noticed that economically valuable seams tended to occur along with certain other kinds of strata and fossils. In addition, it was known that rock strata which disappeared below ground in one region reappeared in other regions and

that their three-dimensional geometry could be measured and worked out. Much of this evidence came from mine workings, especially in Germany where mining engineers and mineralogists such as Lehmann and Füschel drew up some of the first geological maps to show the three-dimensional arrangement of strata.

By the early 19th century, the possibility of making geological surveys of large regions and mapping the distribution of strata on a regional basis was becoming a reality. In France, Georges Cuvier and Alexandre Brongniart (1770–1847) worked out the succession of rock strata in the Paris Basin and drew up a map of their distribution in 1819. Independently, a largely self-taught English surveyor by the name of William Smith (1769–1839) was doing the same thing on an even grander and more ambitious scale. Smith made a geological map of most of England and Wales and parts of Scotland depicting the distribution of the strata at the surface and their underground structure by means of vertical sections – an enormous task for one man to undertake. Smith's map was published in 1819 and was a hugely expensive undertaking that bankrupted him.

Following the publication of these first geological maps, there was something of a race to carve up geological time into recognizable units – systems of related strata with their contained fossils which had been laid down in particular periods. Between 1835 and 1841, two English geologists, Adam Sedgwick (1785–1873) and Roderick Murchison (1792–1871), were responsible for mapping and defining the Cambrian, Silurian, Devonian, and Permian systems and periods of geological time. These divisions are now recognized worldwide and, thanks to painstaking work by generations of geologists, have been subdivided into a hierarchy of smaller units. Therefore, when an American geologist is saying something interesting about a particular geological time span such as the Cretaceous/Tertiary (K/T) boundary in Wyoming, a Danish, Italian, or Japanese geologist knows what is being referred to and can point to the same interval in his or her own region. Only by these means can regional and international geological histories be constructed with any degree of accuracy and confidence.

Such correlation is not always easy, and there are still considerable problems in dating geological events,

Earth formation

In 1644, the French philosopher René Descartes (1596–1650) theorized that the Earth settled out into its different layers as it cooled. Vapours from the inner layers escaped, leaving large caverns into which the crust collapsed, thus creating hollows that filled with water to create seas and oceans and intervening blocks that formed the continents and mountains.

William Smith (1769–1839)

A remarkable innovator in the development of geological mapping, William Smith is a key figure in the history of geology. Born in 1769 in the rural Oxfordshire village of Churchill, as a youth Smith was interested in local rock strata and fossils. He was apprenticed to a land surveyor, and by 1791 had established himself as an independent surveyor in the Somerset coal fields where land owners were keen to find coal buried beneath their land.

The developing canal system was part of the transport infrastructure for handling bulky and heavy materials such as coal and grain. Canal construction and the need for stone and clay required information about local strata and their three-dimensional structure. Smith built up an unrivalled "hands-on" expertise and employed simple but effective techniques for demonstrating his knowledge to potential employers. He collected representative samples of rock strata and their fossils and arranged them in order of deposition, thus building up a generalized geological history of the region.

By 1801 he had travelled over much of southern England and compiled information on the arrangement of its rock strata, which he depicted on a large-scale map giving each main layer a distinctive colour. Smith then embarked on an ambitious scheme to make more detailed geological maps, county by county, which could then be put together to form a countrywide map. By 1819, he independently achieved his goal and published his map entitled *A delineation of the strata of England and Wales with part of Scotland*, on a scale of 5 miles to the inch, accompanied by a memoir listing the strata and fossils by which they might be identified. Unfortunately, Smith became bankrupt and was sent to a debtors' prison. Not until the 1830s was the achievement of this pioneer technician of geological mapping fully recognized by the largely middle-class geological establishment of the day. However, his rehabilitation was rapid and he was soon being lauded as "the Father of English Geology", partly perhaps to try and ward off claims that the French geologists Cuvier and Brongniart were the first to make a geological map of a large area.

processes, and deposits. Even when rocks can be dated, it is important to remember that such measures have error bars — nobody can put hand on heart and say that a particular event or boundary, even the famous K/T boundary, has an exact date such as 65 million years. More accurately, it is 64.98 million years old with an error margin of 0.05 million years.

Just as difficult is the problem of trying to ascertain whether a past event was instantaneous or not. Only certain deposits can be recognized as virtually

instantaneous. For instance, volcanic ashfalls often carry a distinctive chemical "signature" and may be distributed over large regions and, on occasions, globally, within weeks of an eruption. Recognition of such deposits in the geological record is about as instantaneous as we can get.

When it comes to fossils, the appearance of new species in different localities is generally taken as the best indicator of the contemporaneity of the strata. But it has to be realized that it takes a few years for a new species to migrate any distance. There are plenty of living examples of the rapid rates at which new forms can spread, but it still involves a matter of years; nevertheless on the geological scale this is virtually instantaneous.

Right from the early recognition of the main periods of geological time, such as the Carboniferous or Cretaceous periods, it was realized that these divisions represented significant changes in Earth history and the history of its life forms. It was also recognized by the mid-19th century that there seemed to be major phases into which the history of life could be divided, namely the Paleozoic, Mesozoic, and Cenozoic (together known as the Phanerozoic, meaning "evident life"), within which the periods of Earth history are grouped. Older, Precambrian-eon rocks were thought to be completely devoid of life and for a while this eon was known as the Azoic (meaning devoid of life), until the first fossils were found within them. In 1861, John Phillips (1800–74), the nephew of William Smith and professor of geology at Oxford University, drew a diagram showing the fluctuating but overall increasing diversity of life. In this diagram, there were two major downturns: one at the end of the Paleozoic and the other at the end of the Mesozoic.

We are still struggling to understand the exact reasons for these events. The end Paleozoic (Permo-Triassic boundary) event was the biggest, but it is still not clear what caused it. The end Mesozoic (K/T) boundary extinction is connected to a major impact event and climate change, perhaps exacerbated by the major volcanic eruption of the Deccan lavas. The testimony of rocks still has much to reveal, however, as there are some five major extinction events in the history of life which have yet to be explained.

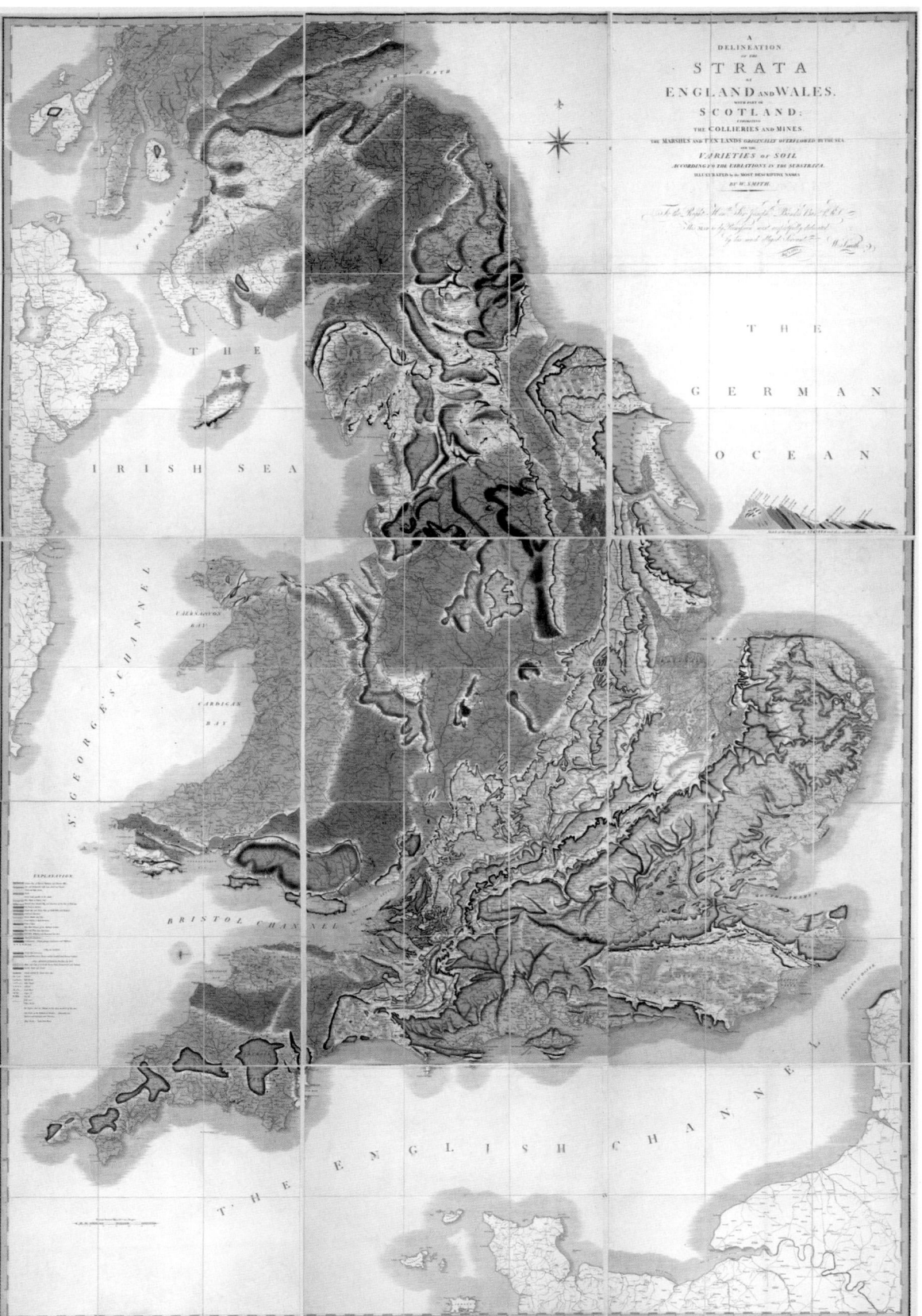

Geological map

In 1819, William Smith's pioneering geological map of the distribution of the strata of England, Wales and part of Scotland was published. It was the first geological map on such a scale to show the three-dimensional overlapping relationship of one strata to another.

Our oldest human relatives

New discoveries in Africa of the remains of our most ancient and remote human relatives are helping to revolutionize ideas about the deep roots of the human family tree. A remarkable skull found in Chad, Central Africa, in 2002 has made a particularly powerful and controversial impact. Named *Sahelanthropus tchadensis* by Michel Brunet, the French leader of the team that found it, the skull is dated at nearly seven million years old and combines a curious mixture of apelike features and more modern human-related ones. The brain is chimp-sized (with a volume of less than 400cc), but the face has a very prominent bony brow ridge, a feature normally associated with much more recent human relatives the australopithecines and early members of our genus *Homo*. Critics claim that *Sahelanthropus* is no more than a fossil ape, but its face is quite different from that of any ape. Certainly its antiquity places *Sahelanthropus* right back at the critical interval when our human ancestors are thought to have diverged from the ancestors of today's surviving apes.

Back in the latter part of the 19th century, Charles Darwin predicted that human ancestors would be found in Africa because the African apes seemed to him to be biologically closest to humans. He was certainly right, as has been verified on a number of counts from comparisons of anatomy, blood proteins, and DNA. We share nearly 50 anatomical features with the chimps, our blood proteins are nearly identical to those of the gorilla, and DNA analysis shows over 98 per cent similarity between us and the chimps, with the gorilla and orang-utan being more distant.

Using the molecular clock, based on known rates of genetic mutation, that estimated two per cent difference between the chimps and humans implies that we must have shared a common ancestor somewhere between 10 and 6 million years ago. This realization caused quite a stir because it had been assumed since the 1950s that our shared ancestry with the apes probably extended 15 or more million years back into early Miocene times. Experts thought that the attributes of humanness must

Hominid fossil skulls
This collection of human-related fossil remains were all discovered in northern Kenya by Richard Leakey in the 1970s and 1980s. From right to left they are *Australopithecus boisei* with its very wide face and small brain; *Homo erectus*, the first human relative to migrate out of Africa, with a much larger brain and heavy brow ridge; and a closely related species called *Homo rudolfensis* with which it coexisted between 2 and 1.5 million years ago.

Dart's discovery

It was in the 1920s that the new picture of human evolution first began to emerge. This was brought about by an Australian anatomist, Raymond Dart (1893–1988), who had survived World War I as an army medic, then trained as an anthropologist in London. In 1922, he was offered an academic post at the University of Witswatersrand in South Africa. Eager to obtain specimens, Dart put out "feelers" wherever he could and was rewarded in 1924 when a small skull was sent to him from some limestone quarries at Taung, near Sterkfontein. Dart described this beautiful juvenile skull in 1925 as a new kind of fossil ape, which he called *Australopithecus africanus*, meaning the southern ape from Africa.

have taken a long time to evolve from a primitive chimplike condition. Fossil apes are known from the Miocene of Africa and Eurasia, but they are all tree-dwelling forms with relatively small brains.

In describing *Australopithecus africanus*, discovered in the 1920s, the anthropologist Raymond Dart claimed that the skull had some human-related features such as small canine teeth and indications that it could walk upright, which was thought to be a fundamental human attribute at the time. To his great disappointment, few other experts agreed with him. Nevertheless, Dart and his friend Robert Broom persisted with their views and continued to find more australopithecine-type specimens.

The 1950s saw the entry of Louis Leakey and his second wife, Mary, into the search for our African ancestors. Leakey chose the Olduvai Gorge within the Tanzanian section of the East African Great Rift Valley because Miocene and younger Pliocene-age animal fossils were known to occur in the region. The Leakeys found plenty of animal fossils and primitive stone tools which were then the oldest known, at around two million years old (subsequently 2.65 million-year-old stone tools were found), but their search for what Louis Leakey thought of as the "missing link" eluded them for a long time.

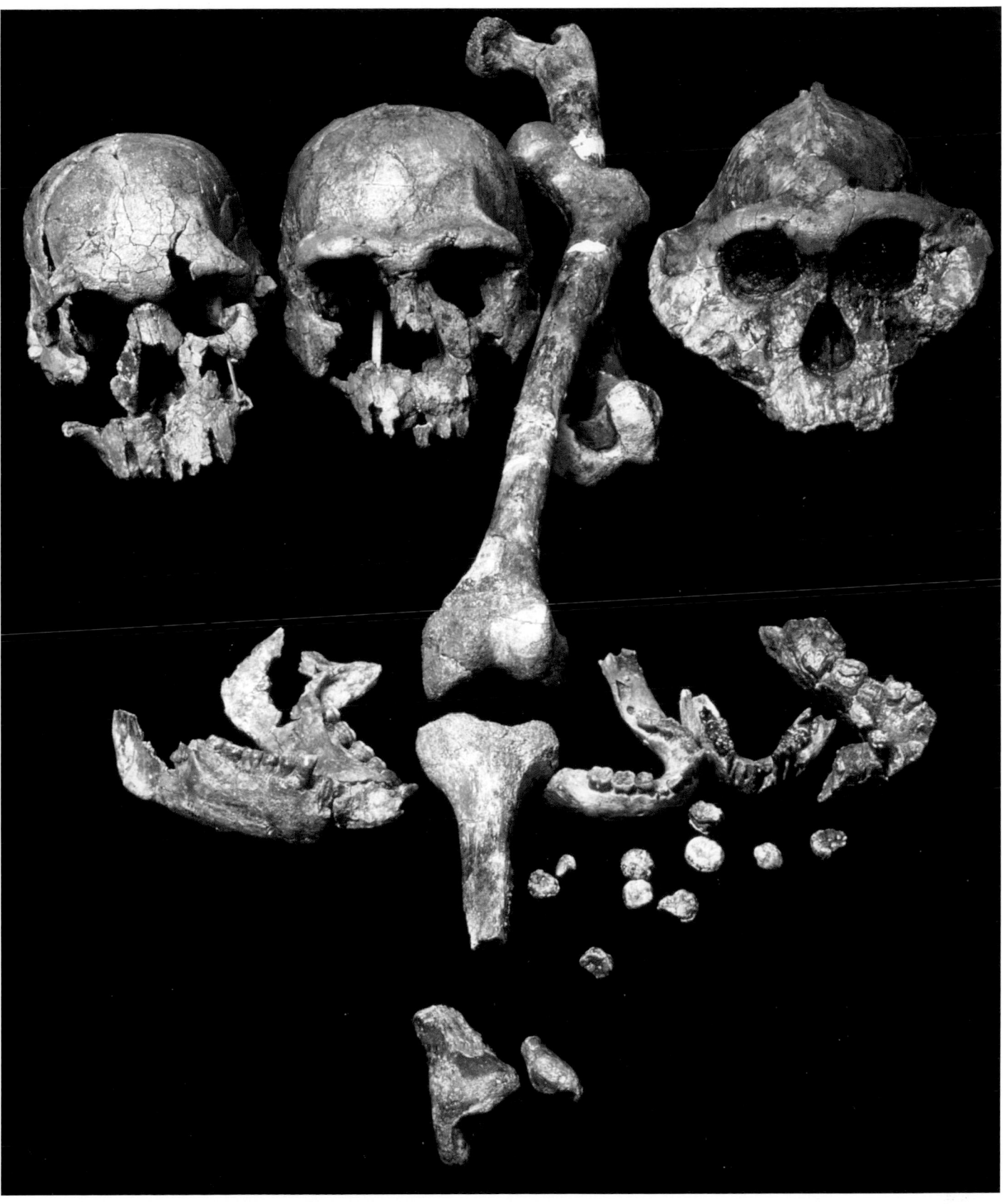

Chimpanzees

Infant chimps learn the skills of tool use through observation and mimicking the behaviour of adults, especially their mothers. Even so, it may take between five and seven years for a youngster to master the technique required to use a hammer stone and anvil to successfully crack open a nut.

Eventually, in 1959, Mary Leakey found a very heavily built skull which Louis named as a new member of the extended human family *Zinjanthropus boisei*, nicknamed "nutcracker man" because of its massive cheek teeth. But it was too small-brained to be seen as a link between the apes and humans. The strata it was found in were dated by the relatively new technique of radiometric dating at 1.75 million years old. The skull was later placed in another genus, *Paranthropus*, along with other robust forms which are now seen as a branch of australopithecines, perhaps related to Dart's Taung find and another spectacular find which was to hit the headlines around the world.

Since the 1960s, many teams of scientists from all over the world have been searching along the Great Rift Valley for human-related remains. An amazing succession of finds has been made, including one of the best ancient human-related skeletons found, the half-complete skeleton of "Lucy" (*Australopithecus afarensis*), found in 1974 by Don Johanson in Ethiopia. Johanson claimed that he had found the "missing link" because, although Lucy was only 1m (3ft) high with an apelike ribcage, her leg bones showed that she walked upright. Bipedalism was still seen as a critical stage in the

transition from ancient apelike relatives to more humanlike ape relatives. Lucy was dated at around 3.18 million years old.

Meanwhile, Mary Leakey and her industrious team discovered a set of remarkable fossil footprints and animal tracks preserved in 3.6-million-year-old ash deposits at Laetoli in Tanzania. Detailed analysis showed that two adults and a child walking close together in a very modern human way had made the human-related footprints. Although no human-related bones were found, the age of the deposit meant that they were probably made by some of Lucy's australopithecine relatives. Clearly bipedalism is very ancient and happened long before there was any great increase in brain size.

Although a number of other australopithecines were found, for a long time, none of them extended back much beyond four million years, and there was still a huge gap in the record of our most ancient relatives. The situation is even worse for the African apes, for whom there is virtually no fossil record until more than 16 million years ago and which comes in the form of *Proconsul*, which combines ape and monkey-like features.

The discovery in the early 1990s of *Australopithecus anamensis* and *Ardipithecus ramidus* took the record back past the four-million-year mark. The *Ardipithecus ramidus* remains are particularly interesting because they combine a mixture of features, including some indications of bipedalism, although perhaps not fully developed, and the teeth have both ape and humanlike features. By placing this ancient relative in a separate genus from *Australopithecus*, the scientists were implying that there may have been more than one stem to the human family tree.

The old view held that certain human characters such as bipedalism, manual dexterity, and large brains all evolved only once in a single-line progression from an apelike ancestor to humans. It now seems that this is unlikely to have been true, and certainly there was no single line of progression, but a much more bushlike or "shrubby" evolution of several kinds of apelike relatives. The new finds of *Sahelanthropus* and another six-million-year-old form called *Orrorin tugenensis*, discovered in 2001, reinforce this view and indicate that we have only just started to lift the lid on human ancestry. There is still much that we do not know about these oldest relatives because the fossil finds are so fragmentary and incomplete.

Sahelanthropus is spectacular and unusual in having a nearly complete skull and lower jaw. The jaw is thicker than that of an ape, and the canine teeth show signs of moving away from an apelike design, but not far. It may be very close to the origin of more humanlike apes. Its more advanced face suggests that some basic human-related features may have evolved more than once. Unfortunately, not much else is known about its skeleton. The discovery of *Sahelanthropus* far outside the East African Great Rift Valley (1500km (930 miles) to the west) suggests that there is plenty of scope for further finds over a much bigger region of the vast African continent. Today, the *Sahelanthropus* find site is a desert; however, around seven million years ago it was forested with lakes, rivers, and abundant game; an ideal setting for our apelike ancestors to live in.

Footprint

This 3.6-million-year-old fossil footprint found at Laetoli, Tanzania was preserved by being covered with warm ash from a nearby volcanic eruption. It was made by an upright-walking human relative, probably *Australopithecus afarensis*, who stood no more than 1.5m (4ft 11in) high.

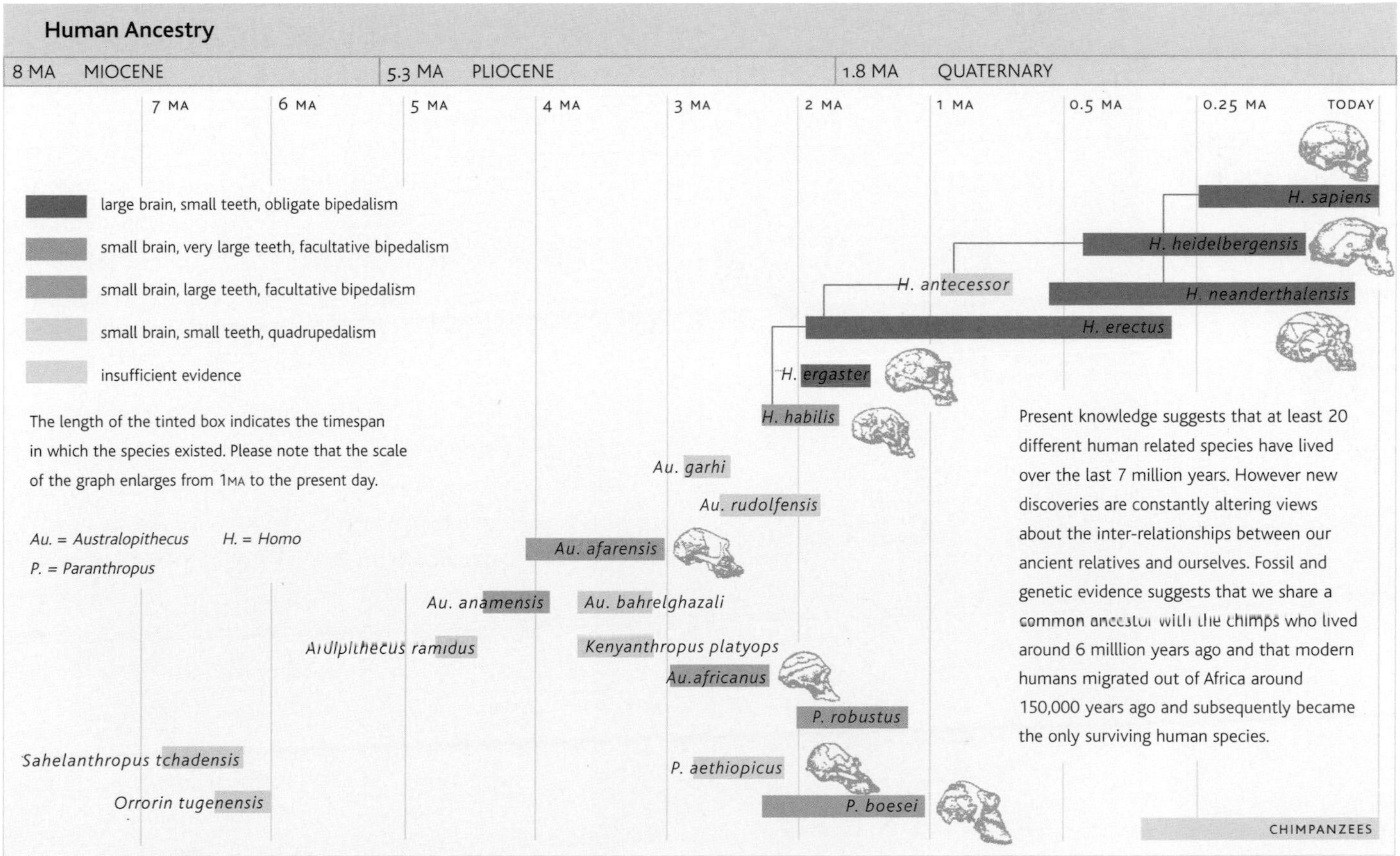

Human Ancestry

8 MA	MIOCENE	5.3 MA	PLIOCENE	1.8 MA	QUATERNARY

| | 7 MA | 6 MA | 5 MA | 4 MA | 3 MA | 2 MA | 1 MA | 0.5 MA | 0.25 MA | TODAY |

- large brain, small teeth, obligate bipedalism
- small brain, very large teeth, facultative bipedalism
- small brain, large teeth, facultative bipedalism
- small brain, small teeth, quadrupedalism
- insufficient evidence

The length of the tinted box indicates the timespan in which the species existed. Please note that the scale of the graph enlarges from 1MA to the present day.

Au. = Australopithecus   H. = Homo
P. = Paranthropus

H. sapiens
H. heidelbergensis
H. antecessor
H. neanderthalensis
H. erectus
H. ergaster
H. habilis
Au. garhi
Au. rudolfensis
Au. afarensis
Au. anamensis   *Au. bahrelghazali*
Ardipithecus ramidus   *Kenyanthropus platyops*
Au.africanus
P. robustus
Sahelanthropus tchadensis
P. aethiopicus
Orrorin tugenensis   *P. boesei*

CHIMPANZEES

Present knowledge suggests that at least 20 different human related species have lived over the last 7 million years. However new discoveries are constantly altering views about the inter-relationships between our ancient relatives and ourselves. Fossil and genetic evidence suggests that we share a common ancestor with the chimps who lived around 6 milllion years ago and that modern humans migrated out of Africa around 150,000 years ago and subsequently became the only surviving human species.

A world dominated by grasses, insects, and mammals

The terrestrial world of Cenozoic times came to be dominated by familiar life forms ranging from flowering plants (angiosperms) to songbirds (neognaths) and mammals. Likewise, the marine world came to be dominated by sea-dwelling mammals from whales (cetaceans) to seals (phocids), along with a myriad of modern bony fish (teleosts) and the more archaic but still eminently successful sharks. The Cenozoic explosive radiation of all these groups (except the sharks) is one of the most remarkable evolutionary features of the past 65 million years and created the modern biological world.

Most Cenozoic life forms can fairly easily be placed in their general categories, such as cats or whales within the mammals, or grasses within the flowering plants. In detail, however, almost all the species are entirely extinct ones, and the further back we go into Cenozoic

times, the less familiar they become. Exotic-looking groups appear in the fossil record and can be difficult for the nonexpert to identify, even at the class level. Animals such as the diprotodontids of Australia, which were cow-sized marsupial mammals, are unfamiliar or indeed unknown. As we shall see, however, the evolutionary roots of these groups extend deeper into the Mesozoic era, when the living world looked very different.

At the base of the food chain on land were the plants, and the Cenozoic plants of the world's landscapes were not so very different from that of today, at least in general terms. Global vegetation zones were broadly similar, although much of the era had warmer climates than the present. Consequently, the tropical, subtropical, and boreal zones were expanded, and the cooler polar and subpolar zones contracted. For instance, in northwestern Europe, palm trees flourished where there are now woodlands of deciduous oaks, beech, and birch – that was, until the climate began to cool around 11 million years ago in late Miocene times.

The dominant Cenozoic trees included both recently evolved flowering ones and more ancient plant groups such as the conifers. Some isolated regions still carried other plant survivors from Mesozoic times such as the cycads, which had been dominant forms, but went into decline in competition with the flowering plants. The great innovation in the plant world that appeared during the Cenozoic occurred when one group of flowering plants evolved into the grasses (Graminae). It is hard to overestimate the importance of this innovation that helped to transform life. An essential component of the success of the flowering plants was what is called the co-evolution of pollinating insects, which was often very specific. Seed- and fruit-eating animals such as birds, bats, and our ancient human relatives also helped the distribution of many plants.

By opening a series of windows, we can look into some of the changes in past life revealed by the fossil record. While the African "Eden" was hosting the

Messel pit plant
This large (23cm (9in) long) fossil leaf with its distinctive venation is typical of the widely distributed aroid plants belonging to the family Araceae, which today grow as lianas or epiphytes, mostly in tropical and subtropical regions.

PERIOD	PALEOGENE			NEOGENE		QUATERNARY	
EPOCH	65 MA PALEOCENE	34.8 MA EOCENE	33.7 MA OLIGOCENE	23.8 MA MIOCENE	5.3 MA PLIOCENE	1.8 MA PLEISTOCENE	10 KA HOLOCENE

evolution of our human ancestors, North America was occupied by a surprising variety of mammals that one might not have expected to have lived on that continent. In late Cenozoic times (the Pliocene, between 5.3 and 1.8 million years ago), with cooling climates, the older forests covering much of the continent were replaced by extensive grasslands. The secret of the success of the grasses lies in their ability to survive and indeed thrive while constantly being cropped by animals, whether domesticated or natural grazers. Unlike many other plants, leaf tip removal does not damage growth because leaf growth originates lower down the stem. Also, many grasses can reproduce from underground

runners, so that they are able to survive even wildfire or severe cropping.

The expansion of grassland as the forests diminished into patchy woodland promoted the expansion and survival of many kinds of grazing mammals, along with the more established browsers such as the elephant-related mastodons. The horses (equids) first evolved in North America around 55 million years ago as small, dog-sized, woodland-dwelling animals; they diversified into many different kinds, increased in size, and became much more fleet-footed to deal with life out on the dangerous open grasslands. However, they, along with so many other medium- to large-sized mammals, became

Snake

This beautifully preserved snake *Palaeopython* is 2m (6ft 6in) long, and was found in the 49-million-year-old Eocene age Messel oil shales, a World Heritage Site near Frankfurt in Germany. The strata preserve a remarkably complete view of an ancient community of plants and animals, from pollen and leaves to insects and fish, birds still with feathers, and primitive mammals with hair.

Primitive horse

No bigger than a large dog, the primitive horse *Propaleotherium hassiacum* is just one of two species found among some 70 specimens excavated from the Eocene-age oil shales of the Messel World Heritage Site in Germany. It has four hooved toes on each of the front feet and three on each of the back. Some of these tapir-like forest-dwelling mammals have been found with their stomach contents of plant-leaf material.

extinct in the Americas during Quaternary times, only to be later reintroduced by Spanish colonists.

There was also a variety of deerlike animals which included some true deer (cervids) and camel relatives. Inevitably, all these plant-eaters attracted the attention of predatory carnivores, including big cats and some hyena-like extinct animals known as borophagines. The

A successful migrant

The continued success of the Virginia opossum (*Didelphis virginiana*) is largely due to its relatively small size (similar to a domestic cat) and its adaptability. It is omnivorous and will scavenge for a wide range of food, can climb well using its prehensile tail, and has plenty of teeth (50) for defence. The downside is that, despite its hair, it is not very tolerant of the cold.

latter had massive, bone-crunching jaws and were probably carrion feeders, operating in groups to drive off the top predators from their kills.

When North and South America were intermittently linked by land, there was an interchange of animals between the two continents. With the break-up of the Pangean supercontinent, South America, like Australia, inherited a primitive mammal fauna dominated by marsupials. Some of these strange creatures took the opportunity to migrate to North America, but today are represented by just one species, the Virginia opossum (*Didelphis virginiana*), which is still remarkably successful. The migration of more modern mammals from the north into South America sounded the death knell for most of the marsupials, who had not, on the whole, had to deal with fast-moving top carnivores such as the big cats. Some of the more modern mammals, such as the giant ground sloths that were abundant in South America, even returned to North America.

Around 30 million years ago (early Oligocene times), the North American "game parks" would have looked much stranger, with a number of large, now extinct mammals grazing and browsing their way across the landscapes. The grasslands were beginning to spread, but there was still plenty of woodlands, which promoted the evolution of a variety of browsers of different sizes that could reach to different heights into the bushes and trees for their food. The brontotheres such as *Brontotherium* were huge, tapir-like animals with rhinoceros-like horns, standing 2.5m (8ft) high at the shoulder. There were also hornless rhinoceroses such as the 4m (13ft) long *Metamynodon* and a number of primitive horse relatives such as *Mesohippus*. In addition, there were some ferocious-looking scavengers such as the hog-like *Archaeotherium* and the bizarre predatory creodont *Hyaenodon*, which may have hunted in packs like the modern hyena.

At the opposite side of the Earth, the last vestige of the Gondwanan supercontinent broke up as the two huge continents of Australia and Antarctica parted, with the Southern Ocean growing between them as Australia moved north. The break-up occurred in late Eocene times around 38 million years ago, and Australia carried with it a remarkable biota of plants and animals from Gondwana, a few of which still survive today. This is especially true of the plants, but also of primitive marsupial mammals who were once much more common the world over. Australia's surviving marsupials, including the kangaroo and koala, are familiar enough today, and there are photographs of the last Tasmanian "tiger", the thylacine marsupial which died in captivity in 1933 and was the last known example of this species in existence. Yet Australasia had a much more diverse marsupial fauna in the Cenozoic era.

Thanks to some remarkable limestone deposits in Australia's northwest Queensland around the town of Riversleigh, a great deal is now known about these animals over a considerable time span from about 15 million years ago. At this time, the region was heavily forested with lowland rainforests, lakes, and rivers developed on a limestone terrain that was riddled with caves, sinkholes, and underground water systems.

Unfortunately, very few plant fossils have survived from these extensive forests because not only did the forest floor soils break down most organic material, but also the soils were not preserved as geological strata, but were themselves eroded away. However, in and around the lakes and sinkholes, carbonate-rich sediment accumulated in hollows, along with the bones of any animals which had drowned in the waters. Carbonate mud can quickly and easily harden into limestone at surface pressures and temperatures. In so doing, it encases and preserves any bone material.

It was only in 1983 that the remarkable fossil faunas of Riversleigh were first discovered, but since then many thousands of almost perfectly preserved bones have been recovered by carefully dissolving the limestone rock in weak acid. Many hundreds of new marsupials and other animals such as snakes, crocodiles, and birds have been found. The marsupials range from cow-sized plant-eaters such as *Neohelos* through flesh-eating kangaroos (*Ekaltadeta*) and the cat-sized predator *Priscileo*, to a variety of other kangaroos and tree-climbing possums such as *Burramys* etc.

With hindsight, no matter how successful the marsupials, they were a sideshow. It was the evolution of the placentals that ensured the mammals subsequent global success. The ability to retain and nourish (via a placenta) a developing embryo within its mother's body, means that at birth the foetus can be much more advanced. Interestingly, even the most advanced placental babies, such as those of deer and cattle which can be mobile within hours, are no more advanced initially than those of turtles or crocodiles. The big difference is made by the post-partum behaviour of the placental mother, who continues to feed her offspring with nutritious breast milk, enabling rapid growth and development. The placental "technology" also allows for the evolution of different strategies of reproduction, foetal development, and care. Numerous small and relatively helpless (altricial) offspring can be produced, or a few bigger and more advanced (precocial) offspring. However, an important variant on the latter strategy occurs where there are a few large but relatively helpless offspring because development has been concentrated in the sensory "apparatus" and its control centre the brain, a strategy adopted by our group – the Primates.

Opening and closing oceans

Rift valley

From a satellite, the rifts forming as plate tectonic mechanisms break Africa away from Arabia can easily be seen extending north from the Red Sea along the Gulf of Aqaba towards the Dead Sea, and along the Gulf of Suez to the Mediterranean, alongside the Nile Delta.

A major discovery of science in the 20th century has been that of plate tectonics. Central to the understanding of its driving mechanism has been exploration of the ocean floor and its geological structure. Following World War II and developments in submarine warfare and technology, the onset of the Cold War in the 1950s added an urgency to the mapping of the ocean floor. The realization that atomic-powered submarines could hide within deep-sea valleys made mapping ocean-floor topography an urgent priority.

The production of the first published global map of the ocean floor by Bruce Heezen and Marie Tharp in 1977 was a remarkable work of synthesis. For the first time, a clear view of the major features of the ocean floor was available. As we have seen, ocean-floor topography is dominated by a series of interconnected rocky mountain ridges. Within the Atlantic Ocean, it is evident that the ridge lies midway between the American continents to the west and Africa and Europe to the east, and it generally parallels the coastal configuration on both sides. If the continents are rejoined along the midocean ridge, there is a near perfect fit along the continental shelf margin. Sampling

of rocks from the ocean floor and from islands flanking the ridge systems shows a similarity of composition in that they are primarily derived from basaltic magmas. Radiometric dating shows that all the ridge rocks are very young and that they become older further away from the ridge. Clearly some mechanism is producing new volcanic rocks along the ridge crest, then carrying them on either side.

The further away from the ridges, the thicker the blanket of ocean-floor sediment becomes until it completely covers the irregular ridge topography. Eventually the sediment builds up into flat-lying, deep abyssal plains some 3–7km (2–4½ miles) below sea level. Drilling cores into this thin sediment blanket reveals that the oldest sediments are found furthest from the ridges, but nowhere is there any sediment older than about 180 million years. It would seem that some process is at work destroying the older ocean floor. It is also evident that in places the ocean floor plunges into even deeper trenches, so deep that some extraordinary force must be holding or pushing the ocean floor down in these zones.

The oceanic ridge system has a number of peculiar features that are fundamentally different from those seen in terrestrial mountain ranges. In cross-section, the ridge is strongly symmetrical, and its crest is frequently offset by transcurrent breaks known as "transform faults". The summit of the ridge is divided by a central fault-bounded rift valley. Large-scale tensional forces must be pulling the ridge apart to produce such rifts. Yet how can tensional forces elevate a huge mountain chain some kilometres above the ocean floor? The answer is provided by the simple physical principle that heated matter expands. Heat-flow measurements show that the rocks beneath the ridge systems are significantly hotter than normal ocean-floor rocks, and calculations show that this is sufficient to expand the rocks upwards to form the ridge. On either side of the ridge, the rocks are cooler and sink deeper down. By contrast, terrestrial mountains have fold and fault structures that are

PERIOD	PALEOGENE			NEOGENE		QUATERNARY	
EPOCH	65 MA PALEOCENE	34.8 MA EOCENE	33.7 MA OLIGOCENE	23.8 MA MIOCENE	5.3 MA PLIOCENE	1.8 MA PLEISTOCENE	10 KA HOLOCENE

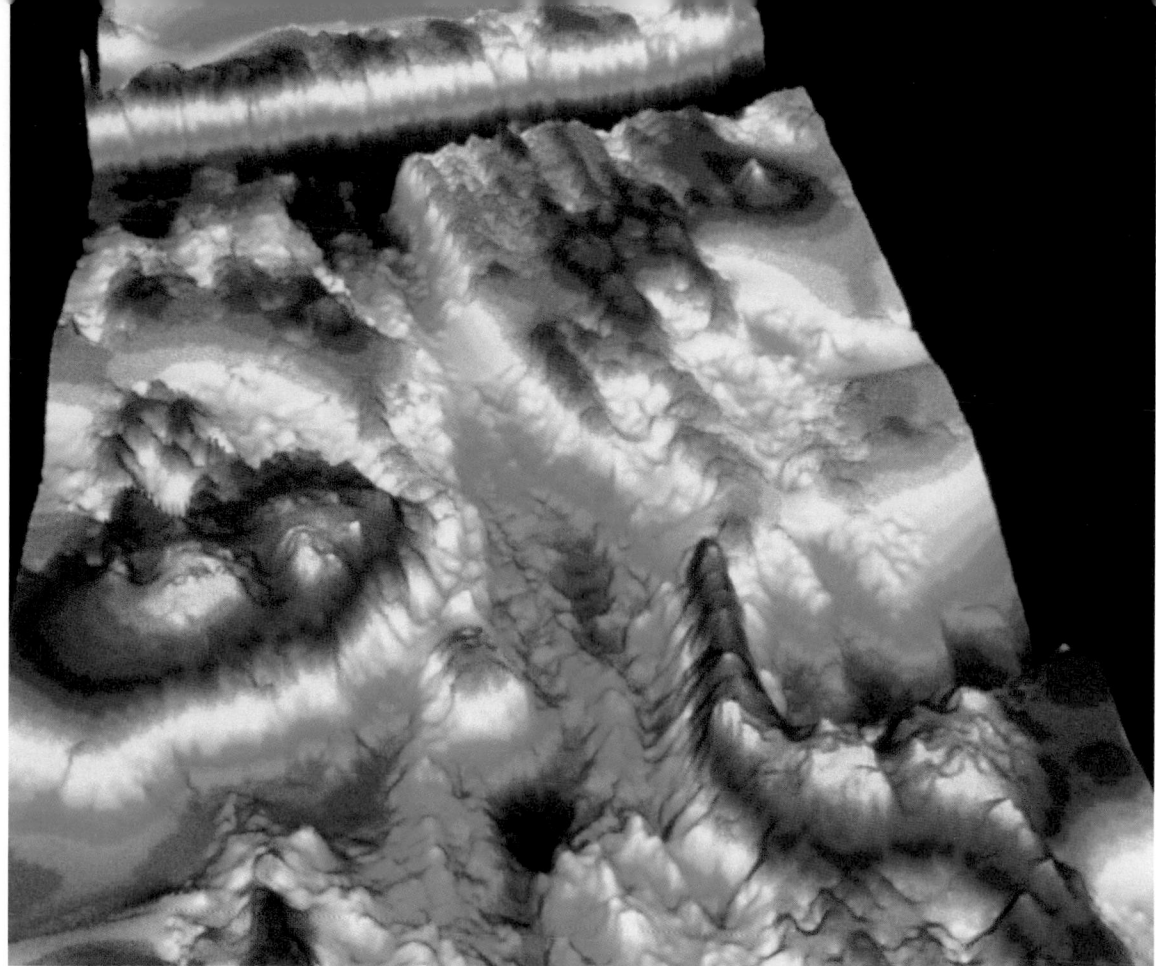

Mid-ocean ridge (left)
Scanned by sonar from a ship, the Mid-Atlantic ocean-spreading ridge runs from top to bottom of this image with its deep central, fault-bounded rift valley coloured blue, and the flanking lava ridges yellow to red, according to elevation. The whole ridge structure is cut by two transform fault valleys running from left to right, one at the top of the picture, the other at the bottom.

essentially the products of large-scale compression and thickening of crustal rocks.

Detailed imaging within the ocean ridge has revealed numerous conelets and linear fissures mark the rift floor, just like volcanic rift valleys on land. Expansion and arching of the ridge rocks generate tension in the crest and the formation of linear cracks (faults). Partial melting of the hot rocks below the crest leads to the upward migration of the melt, which pours out along the fissures as lava. The lava cools, becomes brittle, and is in turn cracked open, with new lava pouring out and so on – a double conveyor-belt system carrying older lava away on either side of the ridge.

The reality of this process of the ocean floor spreading away from the ridges was beautifully confirmed in early 1963 when the Canadian scientist Lawrence Morley saw a new map of symmetrically patterned magnetic anomalies on either side of the Juan de Fuca ridge in the northeast Pacific. It was his "Eureka!" moment because he immediately realized that it supported the theory of ocean-floor spreading and heat convection within the Earth's mantle layer. Like a barcode, these anomalies of ocean-floor lavas recorded reversals in the Earth's magnetic field over time. The fact

that the patterns were mirror images on either side of the ridge and grew older away from the ridge proves that, over millions of years, the ocean-floor rocks are literally pushed away from the ridge by new eruptions. Morley's insight was so innovative that he was unable to get it published; however, later the same year, two British scientists, Fred Vine and Drummond Matthews, came up with the same idea independently and managed to publish. Convection-driven ocean-floor spreading was the driving force behind the generation of new ocean floor.

Black smoker (above)
Hot rocks close to the surface below mid-ocean spreading ridges heat seawater that has penetrated the ridge lavas and recirculates it back to the surface through vents. Minerals scavenged from the rocks are precipitated forming "smokers" and are fed on by extremophile bacteria that survive independently of solar energy.

Building and destroying mountains

Tibetan plateau
Elevated 4–5km (2½ to 3 miles) above sealevel, the Tibetan Plateau, seen here near Lhasa, is the largest high plateau in the world, rising well above the average height of the continents. Coloured Spaceborne Imaging Radar helps reveal the geology of this hilly terrain – granite is orange and brown and older sedimentary and volcanic rock is blue.

Terrestrial mountains are very different from those which rise from the ocean floor. With the average elevation of the land being only a few hundred metres above sea level, a mountain can be defined as any mass of rock elevated significantly above the surrounding landscape. Mountains range from individual volcanic peaks such as Mount Fuji in Japan or Mount St Helens in the northwest United States to ranges which extend across continents and hemispheres such as the Andes of South America. Mountains can vary from sculpted glacial horns such as the Matterhorn in the Swiss Alps to flat-topped elevated mountain plateaux such as South Africa's Table Mountain, as well as the Tibetan plateau and the Altiplano of the Bolivian Andes, with the latter two both rising to more than 5000m (16,400ft).

The major mountains of the world tend to occur within extensive linear ranges which can have very complicated geological histories and structures, such as the Rocky Mountains of North America or the ancient Caledonian mountain belt. The latter stretches from the Scandinavian Arctic Circle, through the northwest of the

British Isles and Ireland, and now, since the opening of the Atlantic continues, to Newfoundland and New England.

Compared with oceanic mountain ranges, those of the continents are much more diverse in origin, age, structure, and rock composition. Some mountain ranges contain very ancient rocks that are thousands of million of years old. But these are unusual – most major mountain ranges such as the Himalayas, Andes, and Rocky Mountains are relatively young structures (a few tens of millions of years old). Nevertheless, they are older and very different in origin from ranges found on the ocean-floor.

Typically, land-based mountain or orogenic belts contain folded and faulted rocks produced by large-scale compression. The same forces also metamorphose rocks at depth, turning muds to slates or schists, and limestones to marbles. Rock melts, known as magmas, form intrusions ranging from vast granite plutons with volumes of many cubic kilometres to thin sheet or wall-like dykes and sills. The latter are generally associated with large volcanic centres and the extrusion of lava and other volcanic products at the surface. Such volcanic centres originate from rising plumes of heat deep within the mantle, which dome and fracture the continental crust with vast outpourings of lava. Some 40 million years ago an event of this kind led to the development of the Brito-Icelandic province and the rifting open of the North Atlantic. Similar processes are still at work in northeast Africa and up into the Red Sea, forming the Great Rift Valley, with volcanoes such as Mount Kilimanjaro and flanking mountain ranges such as the Aberdare Range.

Geophysical surveys confirm that the "roots" of young mountains are much deeper – up to 80km (50 miles) deep – than the elevated portion is high. Powerful compressive forces, resulting from the collision of crustal plates, thicken the crust by folding and faulting. Very large-scale faulting can also thicken the crust by underthrusting large slabs of rock over hundreds of kilometres at low angles. Such processes have probably been responsible for the 5000m (16,400ft) elevation of the Tibetan plateau, north of the Himalayas.

Plate tectonics can explain much of the compressive force that generates linear mountain belts. The formation of the Himalayas is fairly simple compared with more complex orogenic zones such as the Rockies. As part of the Gondwanan supercontinent break-up, the Indian subcontinent split away from Africa around 80 million years ago. The subcontinent was carried northwards as the Indian Ocean grew in size behind it and the Tethys Ocean was subducted in front of it. The subduction process generated volcanoes, and sediments were scraped from the ocean floor, diced, and wedged onto the continental edge, eventually to be caught up in the mountain-building process. Around 40 million years ago, India collided with the southern margin of Asia, and so began the formation of the great Himalayan mountain range.

Where two oceanic plates collide, the subduction of one of them generates volcanicity and the formation of mountainous volcanic island arcs such as the Philippine islands. Large island arc masses resist subduction with the result that they can become part of a growing orogenic zone along with other kinds of plate fragments known as "exotic terranes". There is evidence that the coastal ranges of North America include crustal fragments or exotic terranes that have been transported great distances by plate movements before colliding with the North American plate and being incorporated into the larger orogenic belt.

Detailed studies of the rocks and structures (folds, faults, and other features) of mountain belts both ancient and modern have shown that their development can be exceedingly complex. Now, thanks to the emergence of plate tectonic theory, many of the processes of mountain building can be understood, and the "deep" history of plate movements over geological time reconstructed. The development of mountain ranges now separated by oceans, such as those of the eastern seaboard of North America and northwestern Europe, can be better explained. But many gaps in our knowledge remain especially in relation to the early history of the Earth.

Mississippi Delta

Major rivers such as the Mississippi-Missouri system carry very large loads of sediment, much of which reaches the sea and is deposited in huge deltas which build out well beyond the coastline onto the continental shelf. Such sedimentary environments accumulate large quantites of organic debris, which will form hydrocarbons in the form of coal, gas, and oil in the geological future.

4

Life's middle age

The Mesozoic era or life's middle age is perhaps the best known interval of geological time, since its 173 million years saw the rise and fall of the dinosaurs and encompass the Triassic, Jurassic, and Cretaceous periods. Mesozoic strata are particularly well exposed across much of Europe, having been caught up in the building of the Alps and other young mountain belts across the world such as those in the midwest of North America, South America, China, Mongolia and Antarctica.

Historically, these strata and their fossils have been of enormous importance in the development of our understanding of Earth science. The earliest recognition of the value of Mesozoic strata was made by Neanderthal people over 100,000 years ago when they discovered that Cretaceous age strata over much of Europe contained flint stone which is ideal material for making tools and weapons.

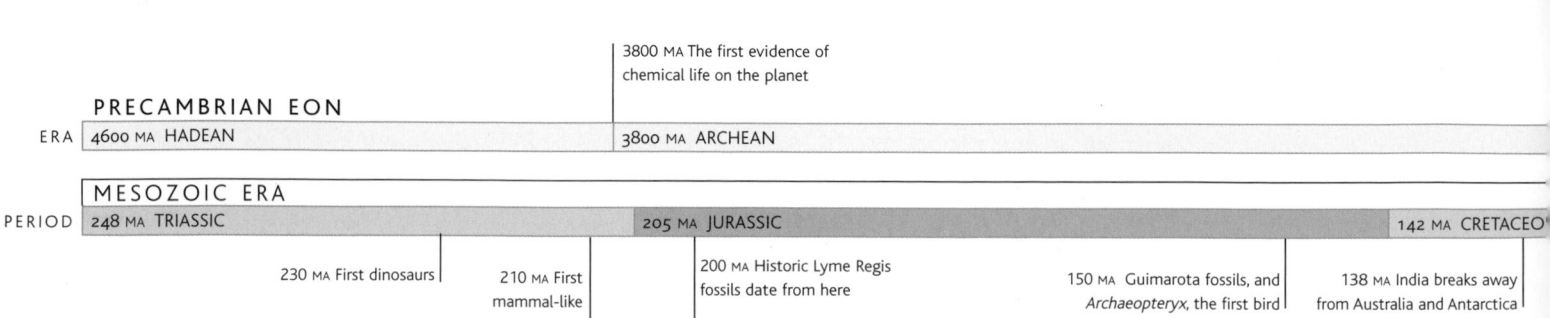

3800 MA The first evidence of
chemical life on the planet

PRECAMBRIAN EON

ERA | 4600 MA HADEAN | 3800 MA ARCHEAN

MESOZOIC ERA

PERIOD | 248 MA TRIASSIC | 205 MA JURASSIC | 142 MA CRETACEO

230 MA First dinosaurs | 210 MA First mammal-like animals | 200 MA Historic Lyme Regis fossils date from here | 150 MA Guimarota fossils, and *Archaeopteryx*, the first bird | 138 MA India breaks away from Australia and Antarctica

200 MA Opening of the central Atlantic and Tethys Oceans

The limestones of the Jurassic age were widely used in Europe for building from pre-Roman times, and it was the growing commercial exploitation of Mesozoic strata in the 18th century that led to the first serious scientific exploration and mapping of their distribution in time and space. But it was not until the 19th century that the now familiar names for the periods within the Mesozoic era were first established.

The discovery of the extraordinary extinct monsters from Mesozoic strata generated great curiosity about their nature and how they fitted into the world order. Firstly, in the latter half of the 18th century, the "Monster of Maastricht" was discovered, followed by a marine reptile *Mosasaurus* from Cretaceous strata, then the Jurassic ichthyosaurs and flying pterosaur reptiles, and finally the dinosaurs during the 19th century. Extinction was clearly a reality, as was the previous existence of a group of animals unlike any now living. These discoveries opened a new window on prehistory which has captivated the public imagination ever since.

Allosaurus

Around 2 tonnes in weight and up to 11m (36ft) long, *Allosaurus* lived in North America in the late Jurassic. The remains of some 40 allosaurs have been found in the Cleveland-Lloyd Dinosaur Quarry in Utah, USA. The first largely complete skeletons were discovered in the late-19th century.

1200 MA The first multi-celled organisms date from the middle of the Proterozoic period

610 MA The first large marine animals appear

PHANEROZOIC EON

2500 MA PROTEROZOIC

545 MA PALEOZOIC

248 MA MESOZOIC

65 MA CENOZOIC

124 MA Feathered dinosaurs and the first placental mammal

115 MA Historic *Iguanodon* fossil dates from here

75 MA *Hadrosaurus* fossil dates from here

67 MA Deccan flood basalts

65 MA The end Cretaceous extinction of dinosaurs, mososaurs, ammonites etc

Evolution

Archaeopteryx

Perhaps the most important fossil ever found (from Solnhofen, Bavaria), the remains of this wonderfully preserved early bird, with clear impressions of its flight feathers, were quickly seen as supporting evidence for the Darwin/Wallace theory of evolution. *Archaeopteryx* combines reptilian and avian features and proved the evolutionary link between the two groups. Now we see birds as feathered dinosaur survivors.

The 1861 discovery in Germany of *Archaeopteryx*, still the oldest bird fossil we have, provided the first good proof from the geological record of the Darwin/Wallace theory of evolution. Initially, the theory was jointly published joint in 1858 under the title *On the tendency of species to form varieties; and on the perpetuation of varieties and species by means of natural selection*. When, in 1859, Darwin expanded his ideas in *On the Origin of Species by Means of Natural Selection, or the Preservation of Favoured Races in the Struggle for Life*, he actually said very little about fossils. He did, however, say a lot about the deficiencies of the fossil record. He knew only too well that there were eminent scientists with a much greater expertise on fossils, such as his ex-Cambridge tutor Adam Sedgewick, who were antipathetic to the whole idea of evolution. Darwin had been forewarned about the kind of hostile reception the theory was likely to receive because, in 1844, the anonymous *Vestiges of Creation*, outlining general ideas about evolution, had been severely criticized.

Darwin's five-year voyage on the *Beagle* from December 1831 to October 1836 had exposed him to the amazing diversity of life in the tropics and the remarkable adaptations of life on ocean islands, which was particularly well demonstrated by the finches of the Galapagos Islands. By 1838, he had read the Reverend Thomas Malthus's seminal *Essay on the Principle of Population* (originally published in 1798) and was formulating the beginnings of a theory of evolution. As secretary of the Geological Society in London, Darwin was at the hub of a buzzing scientific revolution and network of the foremost scientists of the age. He heard all the eminent fossil experts of the time arguing over new fossil discoveries, the nature of the fossil record, and its interpretation.

Was Charles Lyell right in thinking that fossil representatives of most major groups of organisms could be found even in some of the oldest fossiliferous rocks, or was there evidence for some progression in life from the most primitive to more advanced forms? Even

by the 1840s, the jury was still out on this question. In 1842, the anatomist Richard Owen (1804–92) realized that new fossil discoveries showd a very advanced group of reptiles, which he named as the dinosaurs, had become extinct. If the theory of evolution were right, how could such a successful and well-adapted group of animals as the dinosaurs have become extinct? Darwin did not want to become mired in such arguments and stuck to biological evidence to support his theory.

One of the major historic debates of biology was over the fixity of species. The majority of naturalists thought that species could not change in any lasting way, although the famous French biologist Lamarck (1744–1829) had argued that change, promoted by usage, could be passed from generation to generation. For this he was roundly attacked by other biologists, especially his compatriot Georges Cuvier.

Darwin was no supporter of Lamarck's approach, but rather looked to the practical experiences of breeding domestic animals for his ideas. Cattle, sheep, horses, dogs, and some birds (pigeons and fowl) had all been transformed within a few hundred years and a few tens of generations by selective breeding. Could similar selection processes produce such change over time in nature? And what about hybridization? It was well known that crossing a horse and an ass produced a hybrid but sterile mule, which seemed to support the idea of the fixity of species. But there was also evidence that some animals, such as cats or cattle, could be successfully hybridized. Darwin argued that speciation could happen in nature through isolation of populations and adaptation to different environments and biological circumstances. The problem, however, was that the mechanism of such change was not known in Darwin's day.

Richard Owen was one of Darwin's contemporaries and an ambitious man. Owen used his position as keeper of the natural history section of the British Museum to spend the nation's money on important and interesting new fossils and biological specimens. So, when he heard

that a remarkable fossil of an ancient bird had been discovered in a Bavarian lithographic limestone quarry, he moved quickly and bought it for the museum. Of course, as curator he had first bite at this particular scientific cherry, and he carefully described and illustrated its curious mixture of bird and reptile features – feathers and a "wishbone" on one hand and a bony tail and teeth on the other.

When Darwin's staunch supporter Thomas Henry Huxley (1825–95) read Owen's description of *Archaeopteryx*, he saw a wonderful opportunity to promote the theory of evolution and pull Owen down a peg or two. Huxley seized the moment and showed that this bird fossil provides startling evidence for an ancestral link between two classes of animals, the reptiles and the birds. It was just the sort of critical evidence from the fossil record that Darwin despaired of obtaining. It was a matter of pure luck that it arrived on the scientific scene just at such an opportune moment.

Over the past 144 years since the publication of the Darwin/Wallace theory of evolution, exploration of the fossil record has revealed a sample of past life which would have surprised Darwin with its remarkable depth of representation. At the highest taxonomic levels (phyla), most macroscopic organisms have some fossil record, even those which are entirely soft-bodied. The problem arises with the small and microscopic organisms, from the worms downwards, which are entirely soft-bodied, and they, certainly, are severely underrepresented. There are also very particular biases of fossilization towards marine "shell" invertebrates. Nevertheless, by targeting certain rare environments of deposition, scientists have been able to recover some good samples of major groups such as the insects and birds, which normally are not preserved as fossils.

As a result, it is now possible to draw up diagrams showing the geological distribution and relative diversity

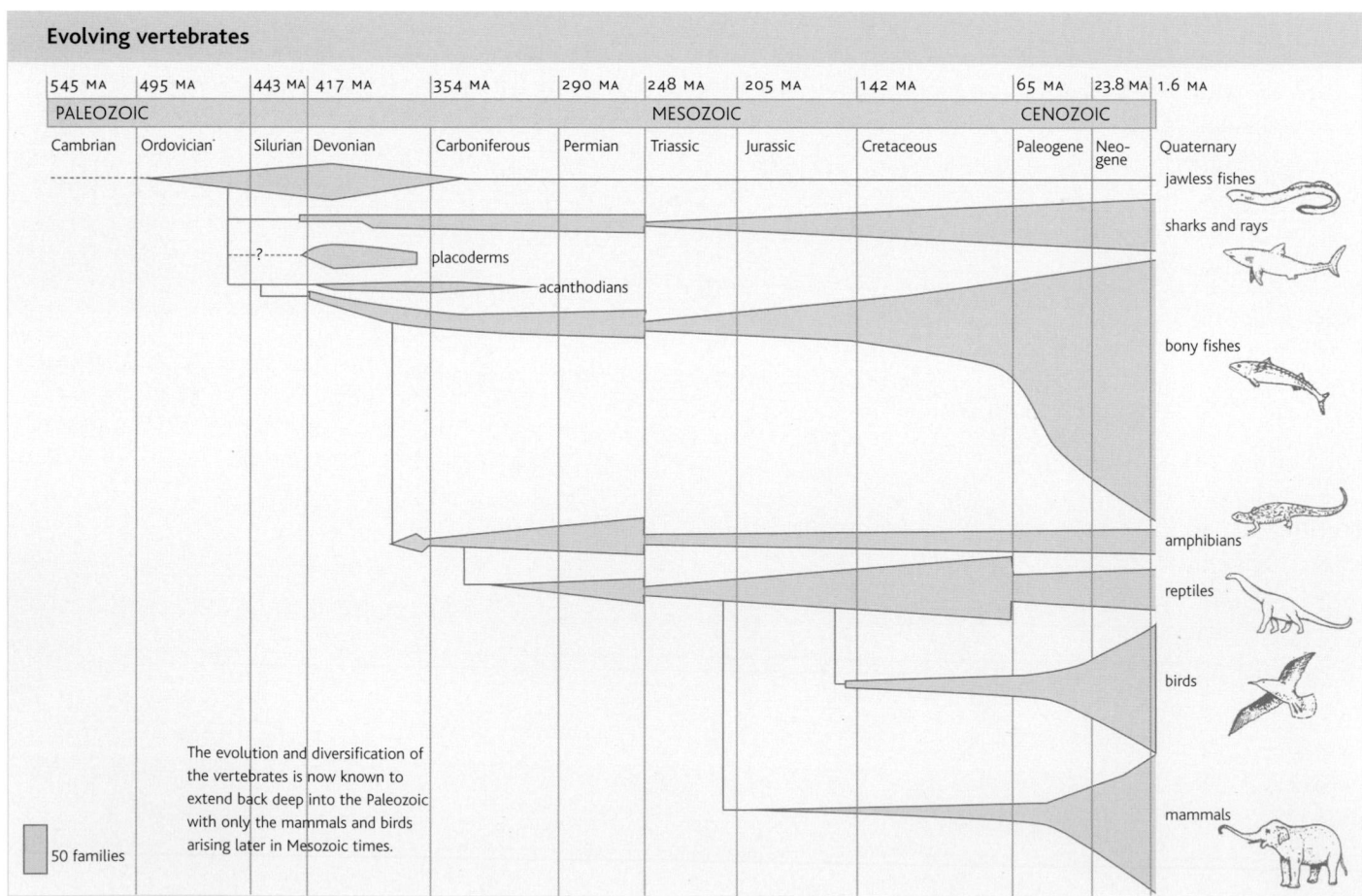

Evolving vertebrates

545 MA	495 MA	443 MA	417 MA	354 MA	290 MA	248 MA	205 MA	142 MA	65 MA	23.8 MA	1.6 MA
PALEOZOIC						MESOZOIC			CENOZOIC		
Cambrian	Ordovician	Silurian	Devonian	Carboniferous	Permian	Triassic	Jurassic	Cretaceous	Paleogene	Neo-gene	Quaternary

jawless fishes

sharks and rays

placoderms

acanthodians

bony fishes

amphibians

reptiles

birds

mammals

The evolution and diversification of the vertebrates is now known to extend back deep into the Paleozoic with only the mammals and birds arising later in Mesozoic times.

50 families

Labyrinthodon — Rhyncosaurus

of the major groups of organisms through time. As long as we are aware of the limitations of the data, diagrams such as this can be useful. This is especially so when it comes to revealing major fluctuations in diversity and significant extinction events. They also show up some interesting problems concerning the origination of the major groups, such as the question of the Cambrian "explosion" of life some 540 million years ago. However, a quite different approach to origination is raising new questions about the completeness of the fossil record.

Not until the rediscovery of the genetic mechanism for heredity in the 1880s and 1890s did the Darwin/Wallace theory of evolution begin to make further headway, although it did have some powerful and very influential advocates such as the famous German biologist Ernst Haeckel (1834–1919). While the Moravian monk Gregor Mendel (1822–84) had, in 1865, published good experimental evidence for inheritance from his work breeding peas, nobody seems to have realized just how important his work had been until it was resurrected by the Dutch botanist Hugo De Vries (1848–1935) in 1899.

Now with the advent of genome mapping, our understanding of the mechanisms of heredity from generation to generation and longer term evolution has made a great leap forwards. Complete mapping of the genome of a number of organsims across the range of life's diversity, from the yeast bacterium to the mouse and humans is indicating levels of genetic similarity and difference across the board. Since James Watson (b.1928) and Francis Crick (b. 1916) first "cracked" the code of life in 1953 with their demonstration of the double-helix structure of DNA, molecular genetics has developed exponentially. Now DNA analysis can prove paternity, convict murderers, and, thanks to the molecular clock, indicate the timing of past evolutionary divergences.

By measuring the genetic distance between primitive representatives of the major groups of living organisms and knowing the average rate of speciation and evolution within the groups, it is possible to estimate when they diverged in the past. Such estimates have been made for most organisms ranging from a divergence time for the ancestors of the chimps and humans, six to ten million years ago, to the origination of the main groups of invertebrate animals around 850 million years ago. Only in the past year or so have fossil finds come anywhere near the ancestry of the chimps and humans. As for the origination of invertebrate phyla some 850 million years ago, the fossil record has a long way to go.

Beasts of the swamp

In retrospect, we can see this as a very fanciful and confused reconstruction (by the 19th-century English artist, Waterhouse Hawkins), of some extinct Mesozoic beasts. At the time, the fossil evidence was very fragmentary and the only available models were living amphibians and reptiles such as lizards. Hawkins depicted the early labyrinthodont tetrapod (centre) as a curious hybrid frog and the reptilian rhynchosaurs as walrus-like.

The discovery of the sea dragons

The first fantastical monsters of life's middle age to be discovered were the sea dragons, and they were found long before the idea of dinosaurs was so much as a twinkle in Richard Owen's eye. The most famous of the early discoveries dates back to the 18th century, when the first fossil remains of what came to be known as the "Monster of Maastricht" were uncovered. Dutch miners found the fossil skull while digging underground tunnels and storage caves beneath the chalk hills surrounding the border town of Maastricht between the Netherlands and Belgium.

The spectacular metre-long fossil jaws became the subject of a long-running argument concerning their zoological affinities. Did they belong to a giant marine crocodile? But they had anatomical differences to crocodiles. Perhaps they belonged to some kind of whale? Scholars from all over Europe came to see them because they also raised the question of extinction. Most still adhered to the idea that a benevolent God would not have allowed extinction to happen to any of his creations. The "monster" was clearly different from any known living creature, but perhaps such monsters still existed in the depths of the oceans? Napoleonic forces besieged Maastricht in 1795, and such was the fame of these "proto-Jaws" that they were seized and shipped off to Paris as war booty. They are still there, and the Dutch have to make do with a plaster cast.

Today, we know that the Maastricht beast was a mosasaur and probably evolved from a group of land-living lizards (the varanids). Complete skeletons up to 10m (33ft) long have since been found and show that they were among the top marine predators of late Cretaceous times, but their success was relatively short-lived (only some 27 million years).

By the first decade of the 19th century, attention had turned to new discoveries being made in Jurassic age strata along the coast of England. Limestones and shales of this age form sea cliffs in both Yorkshire and Dorset. Winter storms undercut the cliffs, causing them to collapse onto the beaches and bringing down great

slabs of rock. Impoverished local people, long used to beachcombing for any saleable flotsam or jetsam, soon found that there was a growing market for fossils among the first tourists to take the seaside air. The locals, often women and children, scoured fresh rock falls looking for fossils shells before the waves pounded them to pieces. Among common fossils such as ammonites, then known as snakestones, they occasionally found bones and even complete skeletons of strange dolphin-like sea creatures.

The skulls had long, toothed, beak-like jaws and large eye sockets, ringed with bones. The paddle-shaped limbs were made of numerous small bones, looked a bit like those of a seal, and were clearly designed for swimming. The last tailbones often appeared to be bent downwards, and so collectors broke them off and straightened them to make the skeleton look better. Scientists who first studied these fossils, such as the Reverend Dr William Buckland of Oxford University, realized that, despite their dolphin-like appearance, these were marine reptiles and quite unlike any living creatures. They were named "ichthyosaurs", meaning "fish lizards", and they were seen as proof that extinction had taken place because their fossil record did not extend beyond Mesozoic times.

Some of the best specimens were found by members of the Anning family in the cliffs of Lyme Regis in southern England. The daughter of the family, Mary, was particularly skilled at finding and preparing specimens for sale, and she became well known in the geological circles of the day. Some of the most famous scientists and even the king of Saxony visited her shop. Although self-taught, she could write informed letters to scientists, such as Professor Adam Sedgwick (1785–1873) of the University of Cambridge, describing new finds. Fossil shells such as ammonites, clams, and ichthyosaurs were not the only things the Annings uncovered. Mary also found a new kind of marine reptile, the long-necked plesiosaur, and some of the earliest fossils of an extinct flying reptile, a pterodactyl, as well as fish and extinct cuttlefish-like creatures known as belemnites.

By the 1830s, the first reconstructions of ancient environments and their inhabitants were attempted. The sea dragons, as they were called, were based on medieval dragon models with serpent-like bodies and tails. All that was missing was St George, the dragon slayer, so they were portrayed as fighting one another in dark, gothic seascapes. More scientific reconstructions took into account the fact that, as reptiles, the ichthyosaurs and plesiosaurs "must" have laid eggs. Consequently, they would have to haul themselves on shore using their paddle-like limbs as seals do. Within a few decades, all this was changed as new finds of ichthyosaurs, especially from Holzmaden in Germany, revealed fully developed embryos within the females, showing that these marine reptiles gave birth to live young in the sea. In addition, preservation of the body outline showed that the ichthyosaur broken tail was not a postmortem artefact, but a distinctive biological feature. The backbone extended down into the lower lobe of the tail and was balanced by a stiffened and symmetrical vertical lobe above. The body and swimming action of the ichthyosaurs turns out to have been like that of fast-swimming tunny fish today. By contrast, the long-necked plesiosaurs used their paddles

Sea monsters

This 1831 reconstruction by the English geologist Henry de la Beche is one of the earliest attempts to recreate a scene from the geological past. It represents life in and around the seas of early Jurassic times some 180 million years ago. The waters are crowded with marine reptiles and their prey, along with seabed shellfish, flying reptiles and the newly discovered land-living reptiles.

Ichthyosaur (right)
A well preserved skeleton of a marine ichthyosaur reptile, found in the 19th century, had its tail straightened by collectors at the time to look better.

Lyme Regis (far right)
Since the early 19th century fossils have been collected from the Jurassic limestones and shales which make up the cliffs around Lyme Regis in southern England. Our modern-day "Mary Anning" ought to be wearing a hard hat because rocks regularly tumble down such cliffs.

in a circular motion to "fly" through the water. Many kinds of these remarkable animals are now known, and they range in time through the Mesozoic from Triassic to late Cretaceous times. They all died out, however, before the end Cretaceous extinction event. In fact, the mosasaurs seem to have replaced the ichthyosaurs as top predators before they, too, became extinct.

Even by the mid-19th century, it was realized that some contemporary groups of sea creatures did die outright at the end of Cretaceous times, notably the molluscan ammonites and their distant relatives the belemnites. The ammonites became especially important because they were so diverse and abundant as fossils. In the 1830s and 1840s, scientists such as the Frenchman Alcide d'Orbigny (1802–57), the Germans Friedrich Quenstedt (1809–89) and Albert Oppel (1831–65), and William Williamson (1816–95) in Britain recognized that fossils could be used to make fine subdivisions of strata. Linear successions of rapidly changing but related species could be used to identify and match successions of strata over large distances. The development of a biozonal scheme based on ammonites was first developed by Quenstedt and Oppel, and is still used in a refined form today. Individual biozones represent periods of time that are as little as one or two million years.

Mary Anning

Mary Anning (1799–1847) was a remarkable woman whose contribution to the scientific study of fossils is only now being fully appreciated. She was one of the few survivors of 10 children born to Mary and Richard Anning of Lyme Regis, a small fishing village in Dorset, England. Impoverished, the family even spent time in the workhouse (1811–15), so out of sheer necessity took to scavenging the local beaches for anything saleable. These were the days of Jane Austen, and the growing moneyed middle class took to visiting quaint villages by the sea. The Anning family sold fossils to these first tourists.

Mary and her older brother Joseph found their first ichthyosaur skull in 1811, and a year later they recovered its body skeleton. The specimen still exists in London's Natural History Museum, but for many years there was no acknowledgement of who had found it. Mary went on to find more important specimens of extinct reptiles, all of which she had to sell. Her surviving letters show that she was making acute observations about the nature of the fossils, and she enjoyed arguing with the experts.

By the late 1830s, however, the fashion for fossil collecting was in decline. Fortunately, the gentlemen of science who had benefited from Mary's finds persuaded the prime minister, Lord Melbourne, to grant her an annual pension of £25 in 1838, but she was probably already ill with the breast cancer which eventually killed her, only five years after her mother's death.

The discovery of the dinosaurs

Today, the common conception of the Mesozoic era is fixated on the dinosaurs of Jurassic Park, as described by the author Michael Crichton and portrayed by Steven Spielberg in his films. *Jurassic Park* is a great story based on the intriguing idea that it was possible to resurrect dinosaurs from the dead through a bit of clever genetic engineering. First, find a mosquito, perfectly preserved in amber and, assuming it has been feeding on dinosaur blood, extract some blood cells and their molecular (DNA) prescription for that particular dinosaur. Next, cut the dinosaur DNA into a reptile or bird egg, so it replaces that of the parent and, hey presto! a dinosaur clone. Unfortunately or fortunately, depending upon your point of view, it is not quite that easy, especially as, despite the reports, no original DNA has ever been extracted and amplified from amber-embedded organisms. DNA is a very fragile molecule that needs very special conditions for its preservation.

Only 200 years ago, it was the extinct marine reptiles, the ichthyosaurs and plesiosaurs, that first grabbed the public imagination because theirs were the first complete skeletons to be recovered and reconstructed. The main reason for this was that most of the Jurassic and Cretaceous strata of Western Europe, where the pioneering work was being carried out, is made up of marine sediments, limestone, mudstone, and sandstone full of the great diversity of fossil organisms which inhabited these warm, shallow waters. There was a scattering of islands, however, and land was never all that far away, although few terrestrial sediments are preserved in this region. Nevertheless, some fragmentary remains of animals and plants which occupied these territories were washed into the nearby seas, where they were preserved to be discovered by geologists in the first few decades of the 19th century.

With hindsight, it is now realized that dinosaur bones were first found many centuries ago in China and even in Europe, but nobody understood what they were. The finds that really set the ball rolling were those of a 25cm (10in) long jawbone with curved, blade-shaped teeth from Jurassic rocks in Oxfordshire, England and some strange leaf-shaped teeth found in Cretaceous strata in Sussex, England. It was William Buckland (1784–1856) who became interested in the Oxford teeth and the question of the kind of animal to which they might have belonged. Clearly, they were those of a large predator and, as they were all of a similar shape, it was more likely that they belonged to a reptile rather than a mammal, which has teeth of various shapes. Buckland named the creature *Megalosaurus*, meaning "giant lizard". But what kind of reptile was it? The only models available at the time were crocodiles and lizards.

The Sussex teeth were radically different and were studied by a local doctor, Gideon Mantell (1790–1852), who was determined to try to make his name as a scientist. Mantell was sure the teeth belonged to some new kind of animal, but did not know what kind, and so he canvassed the opinion of the French anatomist Georges Cuvier. Cuvier at first dismissed them as belonging to some kind of plant-eating mammal. Mantell persisted in his searches and eventually obtained a large slab of rock with a jumble of bones from the same strata from which the teeth had come. There were limb bones and lots of back bones, but unfortunately no skull. Again, Mantell had to struggle to find a model for his reconstruction. Luckily he was shown the pickled body of a marine plant-eating iguana lizard which had recently arrived in London from the Caribbean, and he saw to his amazement that its teeth were remarkably like his fossil ones, only much smaller. Scaling up from the metre-long iguana, Mantell concluded that his fossil animal also had to be a quadrupedal reptile at least 10m (30ft) long and a plant eater, and he named it *Iguanodon* in 1825.

Meanwhile, the brilliant young anatomist Richard Owen (1804–92) was a rising star and one of a new breed of professional museum-based scientists in Britain. He was constantly on the lookout for any opportunity to grab the scientific headlines and get one over

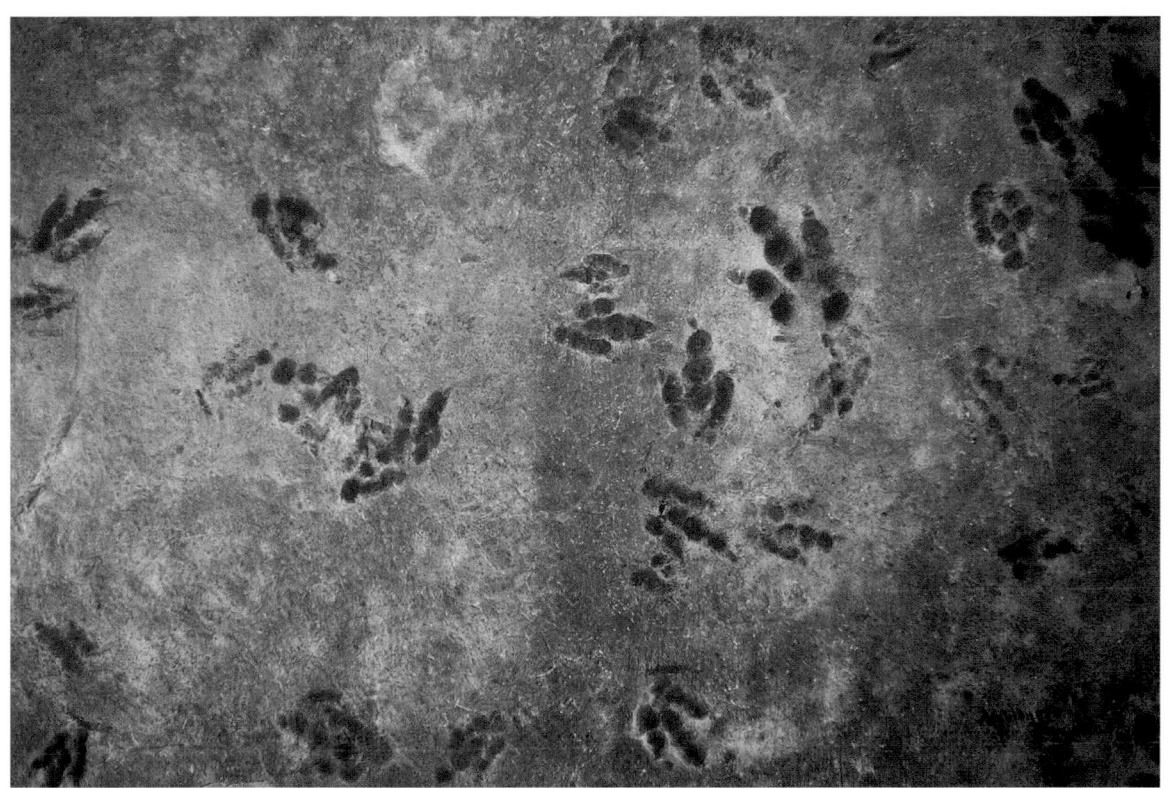

Giant bird footprints

In the mid 19th century, Edward Hitchcock, an American professor at Amherst College in Massachusetts made a detailed collection and study of fossil footprints from Triassic terrestrial strata in the region. Most were three-toed bipedal prints similar to those made by birds only much larger, and Hitchcock's inevitable conclusion was that they must have been made by giant birds. He named and classified the prints according to their detailed appearance (*Brontozoum* in the upper picture and *Gigandipus* in the lower one). Many of his names are still in use, although his giant birds are now known to be dinosaurs and other extinct reptiles and amphibians.

Dino-drama

Dinosaurs have replaced the mediaeval dragon as an icon for monstrous ferocity and danger. The sheer size of many dinosaurs has provided illustrators with an endless source of inspiration for their imagination. Here, in a late Cretaceous scene, two carnivorous tyrannosaurs are depicted attacking a rhinoceros-like, plant-eating ceratopian. In the background, the 10km- (6.2 mile-) wide Chicxulub impactor is seen approaching the Earth and is about to cause the extinction of all the protagonists.

contemporaries such as Buckland and Mantell. Owen had been commissioned to make a particular study of all the newly discovered problematic reptile fossils which were being uncovered from Mesozoic strata in England. In 1841, he presented his review to the British Association for the Advancement of Science. The following year he published his report, which included some important modifications and additions, including the naming of a whole new taxonomic group of extinct reptiles, and so grabbed the limelight from the unsuspecting Buckland and Mantell. Owen coined the name "dinosaur", meaning "terrible lizard", and in doing so opened a Pandora's box of dinosaur mania that still thrives today – if only Owen had registered the name, his descendants would be exceedingly rich.

Although reconstruction of the dinosaurs was constrained by the available reptile models, Owen

realized that such large animals would not have been able to support their massive bodies with crocodile- or lizard-style limbs. When the Crystal Palace from London's Great Exhibition of 1851 was relocated to Sydenham, South London, Owen was given the perfect opportunity to promote his vision of this extraordinary new group of extinct reptiles. Owen tutored some of Queen Victoria's children and had the ear of Prince Albert, who in turn was enormously influential in promoting science and technology in Victorian Britain.

The world's first theme park devoted to prehistoric life, complete with the first life-size dinosaur models, was created by Owen and the artist Benjamin Waterhouse Hawkins (1807–1889) around the rebuilt Crystal Palace. Owen constructed his scale-covered reptilian dinosaurs as curious hybrids with thick mammal-like, elephantine legs supporting massive bodies and heavy, muscular tails which drooped to the ground. Buckland's *Megalosaurus* and Mantell's *Iguanodon* were both seen as huge, cumbersome quadrupeds.

The question of evolution was very much in the air at the time because of the anonymous publication of the bestseller *Vestiges of Creation* in 1844. Owen was highly critical of the work and thought that the evident success of the dinosaurs as a group followed by their extinction argued against ideas of "fitness" and progression. The reopening of the Crystal Palace in 1854 drew thousands of people and so successful was the exhibition that Waterhouse Hawkins was invited to New York to create a similar one for Central Park.

By the time Waterhouse Hawkins arrived in New York, the image of the dinosaur was already changing. By the end of 1858, the substantial part of a new dinosaur called *Hadrosaurus* had been shown by the anatomist Joseph Leidy (1823–91) to belong to a large, bipedal animal. Waterhouse Hawkins set up a workshop and was modelling *Hadrosaurus* when the entire scheme fell foul of local politics and collapsed. *Hadrosaurus* was further transformed in 1866 by American dinosaur expert Edward Drinker Cope (1840–97) into a lithe, leaping, kangaroo-like animal far removed from Owen's imperial leviathans.

Europe still had something to contribute to the *Iguanodon* story. In 1877, the massive bones of some 31 virtually complete members of the genus were found underground by coal miners at Bernissart, in Belgium. They were reconstructed by Professor Louis Dollo using kangaroo and bird models as giant bipedal plant-eaters up to 9m (30ft) long which could use their muscular tails as a kind of prop in order to reach high into tree canopies. It is now known that their tails were stiff, straight, balancing structures that pivoted the body around the pelvis and above the massive hind legs. Fossil tracks show that they were predominantly quadrupedal, but could perhaps rise up on their hind legs for feeding.

By the 1880s, the great American dinosaur race was on, with teams from different museums and universities competing with one another to find bigger and better skeletons in the American Midwest. Othneil Marsh (1831–99) of Yale made the first reconstruction of a

giant plant-eating sauropod 18m (59ft) long, which he called *Brontosaurus*; however, in doing so, he had cobbled together the remains of two different dinosaurs, including *Camarasaurus*, which is the only name now recognized. Not until 1908 was the first of the awesomely large predatory dinosaurs found in late Cretaceous strata in Montana by an expedition from the New York–based American Museum of Natural History. The 15m- (49ft-) long beast was seen as the king of the tyrants of the dinosaur world and given the name that every child over the age of four now knows – *Tyrannosaurus rex*.

Dollo's dinos

One of the most spectacular dinosaur finds in Europe was that of 31 early Cretaceous age *Iguanodon* skeletons in a Belgian coalmine in 1877. The paleontologist Louis Dollo reconstructed some of these 8m- (26ft-) long skeletons as bipeds with a kangaroo-like stance. We now know that they walked on all fours but could rise up on their hind legs to feed on high vegetation.

Life in the shadow of the dinosaurs

The existence of mammals throughout much of Mesozoic times is often overlooked because the dinosaurs receive all the attention and glory. Certainly it is true that for more than 150 million years most land animals more than 1m (3ft) in length were dinosaurs and their reptile relatives. But in the shadow of these remarkable beasts lurked our remote mammalian ancestors – "wee, sleekit, cow'rin, tim'rous beastie[s]", as the 18th-century Scots poet Robert Burns aptly described their modern descendant, the mouse.

These primitive mammals had good reason to be overawed, cautious, and indeed frightened by the reptile-dominated world around them. Nevertheless, throughout much of the time of the dinosaurs, mammals were evolving and diversifying. They were doing so, however, within severe limits. The dinosaurs were not all large and cumbersome – they ranged from pigeon-sized to the famous gigantic sauropod monsters which were pushing the anatomical limits for life on land. A consequence of this diversity was that the dinosaurs and their reptile relatives occupied most ecological niches, and there was not much ecospace left for any other beings.

The few niches that remained were small, dark and crypt-like, such as caves, hollows in trees, and underground. Survival in such limited circumstances places severe constraints on body size and lifestyle. It demanded specially developed sense organs – in other words, good eyes, ears, nose, and some mechanism for feeling the way, such as whiskers. Adaptations of this kind could easily be used for life in the dark in general, ie being nocturnal, especially as reptiles tend to be less active at night, at least in cool or cold climates.

Inevitably, such a hidden or nocturnal life also has to be one of opportunity and carried out either in something of a hurry or very cautiously. The operation of all the sense organs concentrated in the head, where the animal first meets the environment, requires an enlarged brain compared with that of the reptiles. And brains burn energy, requiring constant refuelling, as does the maintenance of a high level of body activity without the warmth provided by the sun. Critical adaptations, then, are warm-bloodedness and an insulating coat of hair.

It used to be thought that warm-bloodedness was a unique attribute of mammals and birds. Over recent decades, however, very good arguments have been presented suggesting that some groups of dinosaurs, especially the small- to medium-sized predators, may well have been warm-blooded too. The discovery of their relatedness to the birds makes this even more likely. Recently, it has been discovered that some of these dinosaurs and their reptile relatives (the flying pterosaurs) had body coverings of modified scales, some of which were hair-like and others which were feather. As they were not used for flight, an insulating function is highly likely.

The constant demand for energy requires food supplies that are preferably high in protein. Insects, other invertebrates, and some plant parts such as seeds are ideal foods, but processing them for rapid digestion and energy release requires a set of specialist tools: teeth. Catching and killing insects requires sharp, dagger-like teeth; crunching hard carapaces or shells requires strong, crushing teeth, preferably with some points; and chopping the food into bite-sized pieces requires ridges or blades – all powered by strong jaw muscles. Tooth differentiation such as this is a particular characteristic of the mammals, although again dinosaur diversity was such that groups such as the oviraptorosaurians (eg *Incisivosaurus*) also evolved a degree of tooth differentiation. There were also important groups of non-dinosaur reptiles that began the trend, particularly the cynodonts of Permian and Triassic times.

The other major advance in adaptation seen in the mammals was in their means of reproduction. Reptiles and birds lay eggs with shells, which is a considerable advance on the unprotected eggs of the amphibians. However, either the eggs have to be big enough to allow

the hatchling to emerge and immediately fend for itself or, if the egg is smaller, the parent has to feed the hatchling until it can do so. Both strategies were and are adopted by reptiles, but eggs and hatchlings are good food and attract the attention of predators. Laying enough eggs to ensure survival of at least some of the hatchlings is not very cost effective. Retaining and feeding the developing embryo in the mother's body until it is more mature can give it a good start in life, but it cuts down on the number of offspring and is a considerable drain on the mother's resources.

Nevertheless, there were enough advantages in the method for it to become the main mode of reproduction in mammals. There are, however, surviving primitive mammals that remind us that this was not always so for the group.

The few egg-laying monotremes of Australia such as the platypus and the echidna — as well as the much greater number of marsupials, with their very underdeveloped offspring kept in pouches after birth — provide good evidence that the more advanced placental mode of reproduction was a relatively late development

Mongolian egg-basket

The first well-preserved dinosaur eggs, nests, and hatchling skeletons were found in Mongolia's Gobi Desert by Roy Chapman Andrews' expedition from the American Museum of Natural History in New York in the 1920s. Since then the region's Cretaceous strata have produced a wealth of fossil dinosaurs and early mammals.

in mammals. The problem for the palaeontologist is that fossilization of the skeleton does not record much information about the niceties of reproduction. Indeed, fossilization tends not to preserve much at all about the small primitive mammals that crept around in the dark while the dinosaurs were not looking.

Small mammal skeletons, as with those of birds, are fragile and easily broken down either by scavengers or microbial processes of decay. The only long-lasting remains are the more resistant teeth, which can pass right through the gut of a scavenger. As most birds do not have teeth, their fossil record is even worse than that of small mammals. Fortunately, mammal teeth are not only preservable, but remarkably diversified in their form and generally distinctive even to the species level. From this tooth-dominated fossil record, several primitive and short-lived fossil mammal groups are now recognized, such as the multituberculates, triconodonts etc.

We now know that from mid-Jurassic into mid-Cretaceous times at least five different and mostly short-lived groups of primitive mammals evolved and died out. There were three surviving groups, one of which was multituberculates, which survived until some 35 million years ago before dying out. The other two are the monotremes and the marsupials, both of which have just about hung on into the present. Right at the end of Cretaceous times, however, a new group of placental mammals evolved. Among this new group of placentals were the earliest primates, small lemur-like animals

Mongolian mammal

Like most early mammals, *Zalambdalestes* was shrew-sized and lived a cryptic existence alongside the dinosaurs and other reptiles of late Cretaceous times in Mongolia. From these somewhat unspectacular beginnings all mammals evolved.

Mining the record of life

The popular view of life on earth in Jurassic times 150 million years ago may be seriously distorted. Tens of thousands of fossils recovered over a period of 10 years by German scientists from the Guimarota coalmine in Portugal record a very different picture from that depicted in the film *Jurassic Park*. Sure, there were plenty of dinosaurs about – 15 different species, identified from some 750 fossils – but none was much bigger than a turkey, and they were outnumbered by lots of other reptiles, especially small lizards, turtles, flying reptiles, and crocodiles. Some of the latter were "salties", marine crocodiles in other words, and by far the biggest beasts around, growing to 9m (29ft) long, substantially bigger than the Australian saltwater crocodiles alive today. The scientists also discovered 9,000 or more fragments of frogs and salamanders, not to mention more than 100 teeth belonging to the earliest bird, *Archaeopteryx*, – the only record of the animal outside of one locality in Germany. All in all, more than enough fossils to keep some 20 experts busy for many years.

Most of the fossils are minute teeth and bones, which have had to be laboriously extracted and then hand-picked from the underground coal deposits. Altogether, the fossils and accompanying sedimentary rocks open a unique window on life in a subtropical coastal swamp just as complex as that of the Florida Everglades today. The biggest surprise is the diversity of mammals, with some 25 species (identified from 7,000 teeth and 800 jaws), again outnumbering the dinosaurs. These distant hairy ancestors of ours were all small, shrew-like and none bigger than a hedgehog. Their delicate bones are not normally preserved, so we usually gain a very biased view from the fossil record which preferentially preserves big bones. Thomas Martin, one of the German team, does not exaggerate when he says: "The Guimarota mine is the most important fossil lagerstatte (accumulation of well-preserved fossils) of the world for late Jurassic mammals and other small terrestrial animals."

which lived up in the trees. A new analysis of the rather poor fossil record we have of the early primates suggests that the last comon ancestor of the primates probably lived as long ago as late Cretaceous times, around 81.5 million years ago. This estimate coincides closely with the divergence times estimated by the molecular clock analysis from living primate groups such as the apes, monkeys, lemurs, tarsiers, lorises, and galagos.

Luckily for us, some of these mammals did manage to survive the Cretaceous/Tertiary extinction event. Within 10 million years, they had become so successful that they split into some 15 different mammal groups, ranging from bats to whales. Even so, six of them subsequently became extinct.

Even as far back as the early decades of the 19th century, before the discovery of the dinosaurs, primitive fossil mammals were found in the same Stonesfield slate strata of Jurassic age which yielded some of the first dinosaur remains. The fossils were small jawbones and

teeth, and, when Georges Cuvier examined them in Oxford University's Ashmolean Museum in 1818, he thought that they belonged to ancient marsupial mammals. He also recognized, however, that they differed from living mammals by having more molar teeth.

One of the specimens was indeed illustrated, described, and named in 1825 as a fossil marsupial, *Thylacotherium*, by the French palaeontologist Constant Prevost (1787–1856). In 1846, Richard Owen showed that it was not a marsupial, but a primitive placental mammal, and he renamed it *Amphitherium*. Nevertheless, one of the other Stonesfield mammal fossils was a genuine marsupial. Even in 1853, Charles Lyell saw these discoveries as "... fatal to the theory of progressive development, or to the notion of the order of precedence in the creation of animals". Within a decade, Lyell had to eat his words and grudgingly acknowledge the concept of evolution.

Insect-eating ancestors

Henkelotherium is a primitive mammal whose well-preserved remains were found in a late Jurassic-age coalmine in Portugal. A woodland species, somewhat like a tree-shrew, it hunted the abundant insects living in the trees whose decayed remains eventually formed coal deposits.

Windows on the past in China

Early flowering plant
A plant from China's amazingly fossiliferous late Cretaceous strata, *Archaefructus liaoningensis*, is thought to be one of the earliest flowering plants (angiosperms), but it still retains a mosaic of more primitive features.

Although the great spread and success of flowering plants took place from Cenozoic times to the present, with some 250,000 species belonging to around 450 families, their ancestry reaches back well into the Mesozoic. Experts on fossil plants (palaeobotanists) have long argued about their early evolution – how and when it happened – because good examples of the delicate reproductive organs (flowers etc) are rarely fossilized. There are some amazingly good late Cretaceous fossil flowers found in Sweden which show that many of the key features were already in place by around 80 million years ago. Part of the problem is that some other more primitive plants, such as the cycad-like bennettitales, also have flower-type reproductive structures, so flowers by themselves do not define true angiosperms. Nevertheless, the early evolution of the angiosperms has to have occurred much earlier, in early Cretaceous or even Jurassic times (some experts claim that angiosperms arose in late Triassic times). The discovery of fossils such as *Archaefructus* with traces of fruits still preserved within the female reproductive structures suggests that there is a good chance that further aspects of early angiosperm evolution will be revealed.

The amazing discovery of dinosaurs with preserved feather-like structures was first met with incredulity and suspicion by some experts. Despite the recent scandal of a forged specimen, there is now no doubt that feathers can no longer be seen as unique to birds because they occur in several small theropod dinosaurs of the early Cretaceous period such as *Sinosauropteryx* and *Caudipteryx*. *Sinosauropteryx*, closely related to *Compsognathus*, was a chicken-sized coelurosaurian covered with downlike feathers and a crest of bristle-like "feathers" running from head to tail. The fossil remains of a lizard and small mammal in its stomach show that it was a very active predator. *Caudipteryx* is a turkey-sized maniraptoran dinosaur which had clusters of better developed feathers attached to its hands and tail. Clearly the evolution of feathers had nothing to do with

Over the past decade, a whole new window to life on land in Jurassic and early Cretaceous times has been opened by a series of stunning discoveries in China and adjacent countries. Several localities scattered across this vast region, from Mongolia and Tibet in the west to Liaoning Province near Beijing in the east have been found to preserve exquisite fossils. These range from one of the earliest flowering plants (angiosperms), the 140-million-year-old *Archaefructus*, and feathered dinosaurs such as *Sinosauropteryx* from the same locality and up to 195 million years old, to early Jurassic primitive mammals such as *Hadrocodium*, from Lufeng in Yünnan Province in southwest China.

flight initially, but became coopted for that purpose in one particular group of small maniraptoran dinosaurs that survived the Cretaceous/Tertiary extinction event and are now recognized as birds.

Chinese and Mongolian localities have also provided very rare complete skeletons of early mammals which give new insights into their evolution and living habits. *Hadrocodium* was a small, shrew-sized animal with a skull just 12mm (½in) long and an estimated body weight of just two grams (0.07 oz). Even more remarkable is the preservation of its minuscule middle ear bones, all three of them, showing that this critical evolutionary feature had already been established by early Jurassic times. In the evolution of the jaw, the original several bony elements of the reptilian jaw are reduced to one (the dentary), as in modern mammals, with the remaining posterior bones forming the middle ear elements and used for improved hearing. The animal also has a bony roof to its mouth (the secondary

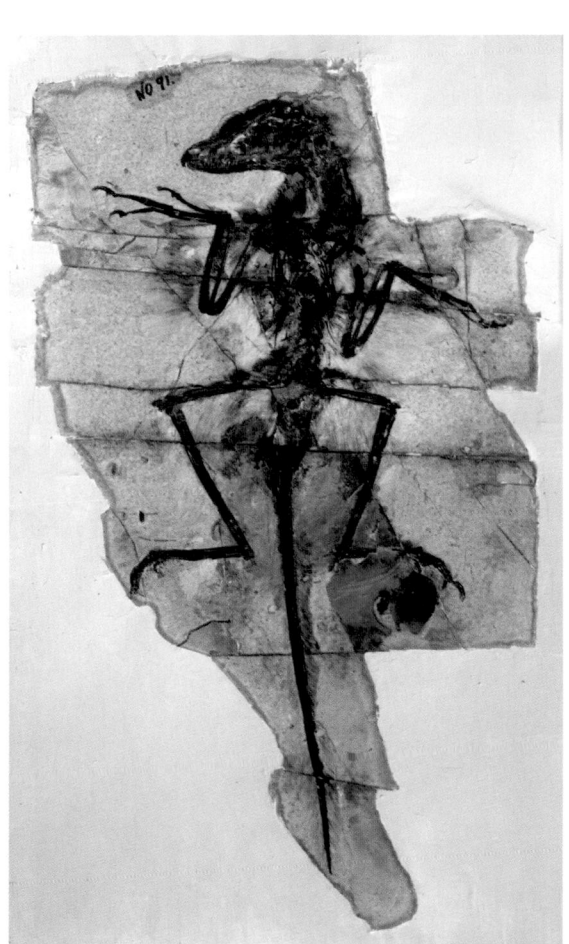

palate), another advanced feature, which allows the animal to breathe and hold food or prey in its mouth at the same time.

Another stunning Chinese find was that of the 125-million-year-old skeleton of *Eomaia*, meaning "dawn mother". Again shrew-sized, this fossil still has traces of its hair preserved in the very fine-grained lakeshore muds that entombed it. Skeletal details indicate that it was a genuine placental mammal, giving birth to live young rather than eggs as in the monotremes, but its pelvic opening was so small the babies must have been very immature and were probably kept in a marsupial-like pouch. The limb bones plus its curved and sideways-flattened claws also suggest that it could climb vegetation as living dormice do. It was accompanied by a larger rat-sized mammal, called *Repenomanus*, which was probably ground living because its fingers and toes were too short to grasp branches. The presence of other mammal fossils belonging to further extinct groups found in the same deposits shows that mammals were already diversified and that their ancestry must lie in yet older rocks. This is supported by the molecular clock, which places the origin of the marsupials at around mid-Jurassic times, some 173 million years ago, plus or minus 12 million years.

Dawn mother (above)
One of the earliest placental mammals is the 14cm- (5in-) long *Eomaia scansoria*, from early Cretaceous strata in China. It was an insect eater with an estimated body weight of 20–25 grams (¾–1oz) and was capable of climbing in bushes and trees.

Fuzzy raptor (left)
Early Cretaceous freshwater deposits in China are providing a wealth of fossils of early birds and small theropod dinosaurs new to science, like the one illustrated here. They have been nicknamed "fuzzy raptors" because of the body covering of fur-like down. Despite the longer arm feathers they could not fly.

5

Life's extinction events

Ideas about the nature of change in prehistoric life have vacillated over the centuries between the sudden and the gradual. The classical western and Judeo-Christian view tended towards the sudden and catastrophic. God was wrathful and capable of both creating and destroying life in an instant. The Old Testament saw life created within days, and virtually wiped out by the Noachian Flood.

Early scientific views of the prehistory of the Earth and life tended to build on this catastrophic vision. The succession of sedimentary rock strata with their fossils was thought to result from one or more catastrophic floods. However, by the beginning of the 19th century, naturalists who explored the fossil record found that it extended much deeper into the stratigraphic rock record than previously expected. Furthermore, fossil representatives of the major groups of living organisms continued to be found in older and older strata.

3800 MA First evidence of
chemical life on the planet

PRECAMBRIAN EON

ERA	4600 MA HADEAN	3800 MA ARCHEAN

PALEOZOIC ERA

PERIOD	545 MA CAMBRIAN	495 MA ORDOVICIAN	443 MA SILURIAN	417 MA DEVONIAN	354 MA CARBONIFEROUS	29

443 MA End Ordovician
extinction, 53% marine genera extinct

450 MA Late Ordovician glaciation

The Scottish geologist Charles Lyell, who exercised a powerful influence on 19th-century scientific thought, even suspected that fossils of all groups would be traced back to some creative moment in the remote past. But he soon had to abandon that position as it became clear that there was some sort of progression from primitive to more advanced life forms through geological time. This latter view was supported by the Darwin/Wallace theory of evolution, which was essentially gradualist and required small changes over a very great deal of time.

At the same time, paleontologists were finding that the distribution of life through geological time was not as smooth as Darwin expected it to be. Even by the 1860s, John Phillips, William Smith's nephew showed that the three great eras of life, the Paleozoic, Mesozoic, and Cenozoic were separated by significant declines in diversity which are now recognized as major extinction events. It now seems that there is both gradual and catastrophic change, the question is, which is the more significant?

Precambrian Ice Age

This mixture of irregularly shaped boulders scattered through finer grained deposits in Varangerfjord, Norway is typical of a fossil till deposit generated by glacial erosion and deposition. The scratches on the rock pavement in the foreground are evidence for glaciation.

2500 MA "Snowball" Earth glaciation	1200 MA The first multi-celled organisms date from the middle of the Proterozoic period		700–580 MA "Snowball" Earth glaciation	610 MA First large marine animals appear	575 MA Abundant Ediacarans	

PHANEROZOIC EON

2500 MA PROTEROZOIC	545 MA PALEOZOIC	248 MA MESOZOIC	65 MA CENOZOIC

MESOZOIC ERA

CENOZOIC ERA

...MIAN	248 MA TRIASSIC	205 MA JURASSIC	142 MA CRETACEOUS	65 MA PALEOGENE	23.8 MA NEOGENE
	248 MA Permo-Triassic extinction event, globally low sea-levels, 68% marine genera extinct	200 MA Eastern North American plateau basalts	133 MA Parana (South America) and Etendeka (SW Africa) basalts	65 MA Chicxulub impact event, 43% of marine genera extinct	
	250 MA Siberian flood basalts	200 MA opening of Central Atlantic		65 MA Deccan lavas	
		205 MA End Triassic extinction, 45% marine genera extinct	133 MA Opening of South Atlantic		

The discovery of extinction events

The idea and experience of catastrophic change in the history of life and Earth's environments are not new. Indeed, accounts of such changes are among the most ancient in human history and prehistory, and have entered the folklore of many human populations around the world.

Most people experience catastrophic events at some time during their lives, but some people experience these events more frequently than others, depending on where they live. Storms, floods, tsunamis, earthquakes, and volcanic eruptions take place all the time, but their frequency, strength, and effects on life vary enormously, and they are very unequally distributed around the world. Storms and floods tend to be more catastrophic in tropical regions, where tropical cyclones occur. Earthquakes and volcanoes tend to occur at the boundaries between crustal plates. Tsunamis hit coastal locations adjacent to submarine seismic events such as earthquakes or large-scale submarine slumps. But "adjacent" in this case may be on the other side of an ocean because tsunamis can cross vast bodies of water without losing their destructive power.

Most of these catastrophic events may be disastrous for life in the immediate vicinity, and indeed both earthquakes and floods have killed huge numbers of people. For instance, the 1976 Tangshan quake in China caused some 242,000 fatalities, while the Hwang-Ho flood of 1887 drowned some 900,000 people and large numbers of livestock. At the human scale, these events are catastrophic events. In geological terms, however, these are still local events in the extent of their impact and have little long-term effect. Indeed, there may be no evidence whatsoever of them in the geological record.

Nevertheless, the geological rock record does store evidence of a variety of much larger catastrophic events. Fortunately, they are the rarer catastrophic events that happen so infrequently that they have not occurred within the seven or eight millennia of recorded history. The rock record also reveals instances of relatively slow but still catastrophic change in climate and sea level on a global scale, and these have all been known about for more than 150 years. The past few decades, however, have seen the discovery of equally disastrous but much more rapid catastrophes, such as extraterrestrial impacts which seem to have played a major role in changing the course of life on Earth.

With hindsight, we can now understand why flood stories have played such a significant role in the history of many ancient peoples around the world. These were very real events which would have had catastrophic impacts on regional scales. Human populations were much smaller in prehistoric times, and so regional floods would have been devastating. We now know that early modern humans moved out of Africa around 100,000 years ago, perhaps pushed by climate change. Those that moved north into Asia and Europe encountered a world of ice ages, rapid climate changes, rising and falling sea levels, and frequent flood events at times of glacial melting. And, as we now know, these very efficient bands of hunters unwittingly created their own catastrophic impact on the animals they relied upon for food, instigating an extinction event which continues today.

Early interpretations of the fossil record portrayed the entombment of shells and bones in strata as a result of the Noachian Flood. This was subsequently modified to include a succession of flood events when it was realized that fossils occur throughout great thicknesses of strata that could not all be the result of a single event. Georges Cuvier, the famous French naturalist, firmly believed that a series of catastrophes was responsible for the fossil record. However, by the early decades of the 19th century, most earth scientists regarded the fossil and stratigraphical rock record as a sequence of gradual depositional events.

What they also began to realize was that something strange had occurred at various times in the succession of fossil life. When scientists were first mapping out the succession of the world's strata and subdividing geological time, they were using the distribution of fossils as a basis for making their subdivisions. It was in 1840 that William Smith's nephew, John Phillips, first pointed out two very marked changes in the geological history of life. The first break occurred at the end of the Permian period and the second at the end of the Cretaceous period. He used the term "Paleozoic" (Greek for "ancient life") for the era up to the end of the Permian; "Mesozoic" (meaning "middle life") for the Triassic, Jurassic, and Cretaceous periods; and "Kainozoic" (subsequently modified to "Cenozoic", meaning "recent life") for post-Cretaceous time.

The main justification for these three great eras in life's history was the recognition of major breaks in the succession of life forms. As Phillips wrote in 1841, "There are whole groups of organic forms which occur only in the oldest strata, others which prevail only in the middle, and some which are confined to the upper deposits." Trilobites, corals, scaphopods, rostroconch bivalves, hyolithids, and blastozoan echinoderms are examples of marine invertebrate groups which became extinct at the end of the Paleozoic. Dinosaurs and flying pterosaurs reptiles are examples of vertebrate groups which became extinct at the end of the Mesozoic, and ammonites and belemnites are invertebrate marine examples.

The most famous of these changes in life is the termination of the reign of the dinosaurs at the end of the Cretaceous period; however, the likely cause for their untimely end has only recently been recognized. By comparison, the end Paleozoic event was even more devastating in its global effect, but the exact cause of this event is still elusive.

A distinction has to be made between catastrophic extinction and "normal", or background, extinction. No

Italian boundary layer
A thin black clay layer (2cm (³/₄in) thick, with the coin for scale) marks the boundary between an older Cretaceous limestone and an overlying younger Cenozoic (Tertiary) one at Gubbio in northern Italy. Geochemical analysis of the clay shows a "spiked" increase of iridium levels indicating the sudden introduction, 65 million years ago, of a relatively large amount of this rare, extra-terrestrially derived element.

Danish boundary layer
The Cretaceous/Tertiary boundary with the iridium spike is seen in many localities around the world. Indeed, its global occurrence is evidence of the magnitude of the impact event. Here, for instance, at Stevns Klint on the Danish coast, the boundary is marked by a clay full of fish remains separating uppermost Cretaceous chalk from earliest Tertiary chalk.

species lasts indefinitely, but either dies out or evolves into one or more new species, and this process is an ongoing one. Typically, species only persist for, at most, a few million years. Human-related species such as the Neanderthals (*Homo neanderthalensis*) only survived for some 300,000 years before becoming extinct without evolving into a new species, although at one time it was thought that they may have contributed to the modern European human gene pool. (DNA evidence has subsequently disproved this.) Nevertheless, the extinction of the Neanderthals may well be seen as part of a wider but relatively minor extinction event, associated with the latter part of the Quaternary ice

ages and the destructive spread of modern humans. Although this event has been spread over 50,000 years and more, on the geological time scale, this is still considered a rapid event.

In the last few decades of the 20th century, serious efforts were made to collate and quantify changes in fossil life through time. American palaeontologists Jim Valentine and J. John Sepkoski Jr led the way. The vast, scattered scientific literature, written in many languages over the past century and a half, was scoured for information about the numbers of fossil taxa, their distribution, and their duration throughout geological time. This was attempted at different taxonomic levels to see what patterns emerged, but it was soon realized that there are many problems with this kind of data set. For instance, many taxa are duplicated because different scientists have given them different names. The accuracy with which their stratigraphic ranges are given varies enormously and tends to refer to the nearest known boundary, so that inevitably there is a false clustering at boundary intervals. Certain geological strata, especially the younger ones, have been investigated to a greater extent than older strata, in what is known as the "pull of the recent". Some geographical areas, such as Europe, have been investigated in more detail than other parts of the world. Nevertheless, it was still an extremely useful exercise and drew a picture of the diversification of life over time which was a significant advance on Phillips's 19th-century version.

The distribution curve was based on families of marine creatures because theirs is the best and most consistent record in rock strata. The pattern showed an initial steep increase in diversity from late Precambrian times, accelerating through the Ordovician only to drop dramatically at the end of the period. There was a recovery through the Silurian, but the overall diversity reached a plateau and remained there throughout the rest of the Paleozoic, terminated by another dramatic fall at the end of the Permian, just as Phillips had shown more than 100 years previously. Recovery during the ensuing Triassic was interrupted by another sharp fall at the end of the Triassic, but thereafter there was a steep recovery through the rest of the Mesozoic era, with total diversity finally climbing above that of Paleozoic times during the Jurassic. But then the Cretaceous is terminated by yet another sharp fall, followed by a rapid recovery and another acceleration through early Cenozoic times which has slowed, but still climbs to the present.

Here was graphic evidence of at least four major prehistoric setbacks to the evolution and diversification of life. These seemed to be clear extinction events, but were they real and, if so, what had caused them?

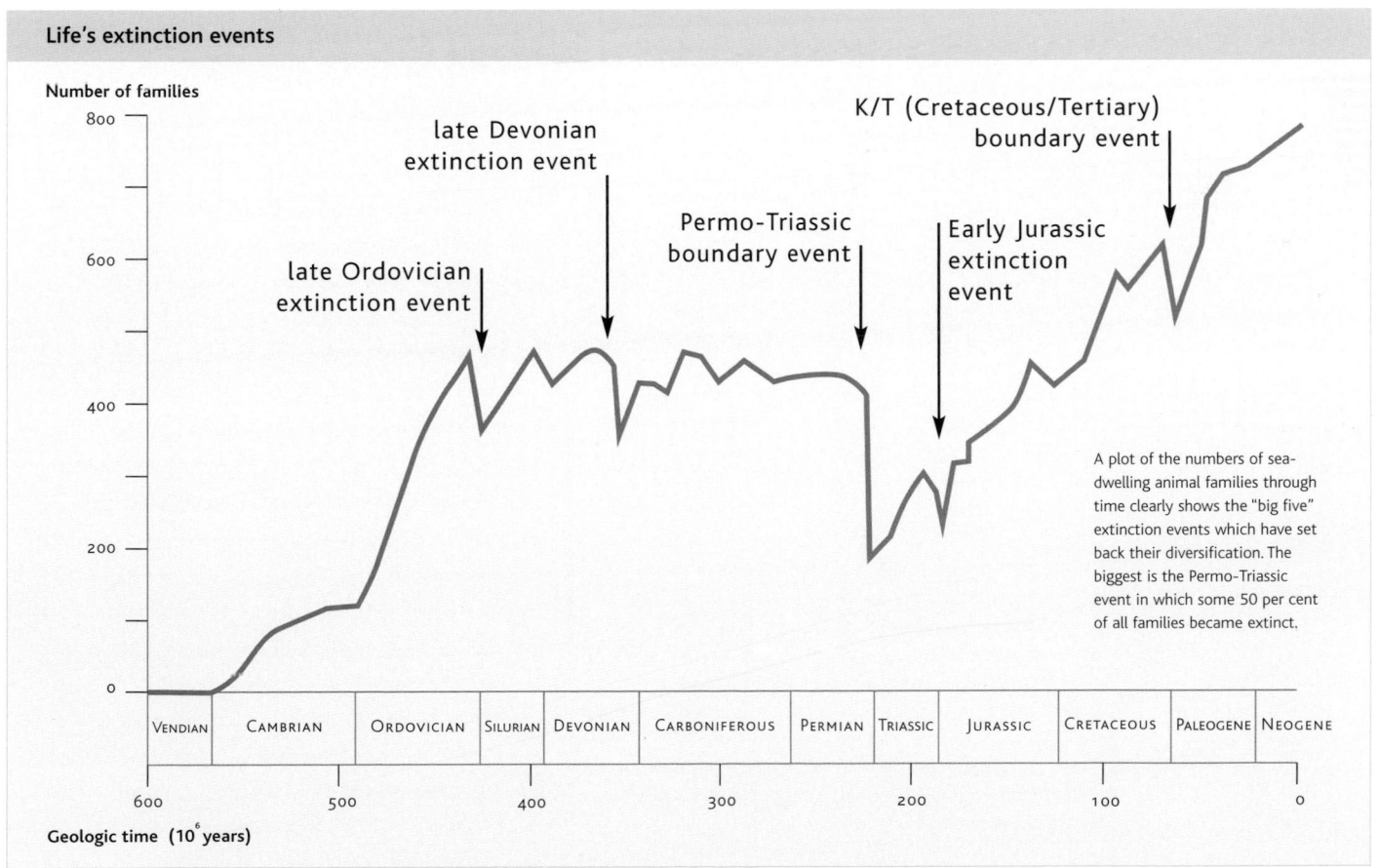

Life's extinction events

A plot of the numbers of sea-dwelling animal families through time clearly shows the "big five" extinction events which have set back their diversification. The biggest is the Permo-Triassic event in which some 50 per cent of all families became extinct.

Extraterrestrial events

By the 1980s, the extinction event that took place at the end of the Cretaceous period was a well-known phenomenon, in particular the demise of the dinosaurs. Many theories had been advanced to explain it, ranging from virus pandemics to climate change stimulated by very large-scale volcanism. The latter seemed a very real possibility because the one big event known to have occurred at this time was the outpouring of the Deccan plateau basalts in western central India.

Arguments based on why certain fossil groups became extinct and others did not were not helping to clarify what the cause might have been. Scientists from different disciplines were employing their own expertise and techniques to attack such problems. Some of the best sections through the Cretaceous/Cenozoic boundary, known generally as the K/T boundary, are in seabed sediments, especially those found in Europe.

The problem is that, because the Earth is still fortunately a dynamic planet, the surface is constantly being reworked by a variety of geological processes which are remarkably good at removing even the largest natural phenomena such as mountain ranges and large-impact craters. Even so, to have caused such an extinction, the K/T impactor must have been big and have left an even bigger hole, perhaps in the order of a 100km (60 miles) or so wide. As it happened only 65 million years ago, there should be some signs left on land, but what if it had landed in the ocean? With the more rapid reworking of ocean floor material, it was quite possible that the site would never be found because it could have been subducted. And even if it had not, it could only be "spotted" on the deep ocean floor by some indirect means such as a geophysical survey. Such was the interest in the story, however, especially because of the newsworthiness of the dinosaur connection, that lots of scientists were very keen to receive the credit for finding the impact site.

Meanwhile, other geologists were tackling different aspects of the event on land. They soon found that the iridium "spike" anomaly could be picked up in other sections of the K/T boundary around the world. In addition, close examination of the boundary sections turned up another strange sign of the true nature of the event: shocked quartz grains. Tiny, sand-sized particles of the common mineral quartz were found with sets of parallel fracture planes running through them that are not normally seen in this mineral. Experiments showed that they could only be produced by very high-pressure shock waves, even higher than those induced by volcanic explosions. It was also realized that the relative abundance of shocked quartz was likely

The K/T iridium anomaly

During the 1960s, Italian experts on microfossils were working on a particularly good section near the historic Tuscan town of Gubbio. The strata here show a marked change in the kinds of microscopic foraminiferan fossils found immediately below and above the boundary, which is itself marked by a thin non-fossiliferous clay layer. An American geochemist by the name of Walter Alvarez sampled the clay layer as part of an analytical survey and was surprised to find that the clay contained a very rare element called iridium. Although it was only present in very minute quantities (a few parts per billion), iridium is normally much rarer than this in the Earth's crust. In fact, it is only known to occur in the Earth's core and in extraterrestrial rocks such as meteorites left over from the early formation of the Earth.

Walter Alvarez's father happened to be the Nobel Prize–winning physicist Luis Alvarez, so naturally Walter discussed this strange iridium anomaly with his father, who took no time in piecing the puzzle together. The only way the iridium could end up in a deep marine clay was from dust introduced by the impact of a very large body from outer space. Walter knew very well that one of the most famous extinctions to occur at the end of the Cretaceous was that of the dinosaurs. In 1980, Alvarez father and son (along with Frank Asaro and Helen Michel) published their ground-breaking speculation that it was just such an impact event that had caused the extinction of the dinosaurs and so much else. The question was, if there had been such a globally catastrophic event, where was the big hole that would have been caused by it?

to be higher nearer the impact site, and soon mapping of shocked quartz distribution seemed to be pointing towards the general region of the Gulf of Mexico and the Caribbean.

In retrospect, the site of the impact was found in the early 1980s by Mexican geologists doing exploratory drilling for the Mexican state oil company. They even published a short synopsis of their discovery in a major American geological journal, but nobody else realized the significance of what they had found at the time. Then, in 1991, a geophysical survey picked up a whole series of anomalies over the Yucatán peninsula in Mexico and out into the Gulf of Mexico. These showed a large, hidden, multi-ringed structure blanketed by a kilometre of more recent sediments. The diameter of the furthest ring was more than 200km (120 miles), and there was a clear high in the middle of the structure. The site of the K/T

impact had been found and became known as the Chicxulub crater, after the nearest town in the peninsula.

The discovery of the impact site coincident with the K/T boundary extinction event reinvigorated investigation of the whole phenomenon. How could an impact event, however big, in the Gulf of Mexico cause a global extinction affecting life in both the sea and on land?

The amount of iridium found in the boundary spike suggested that the impactor had to be an asteroid or comet some 10km (6 miles) in diameter, which would have been vaporized by the energy of the impact. It was thought that the dust thrown into the atmosphere would have caused a scenario similar to a "nuclear winter", with catastrophic loss of sunlight, atmospheric cooling, and a collapse of plant life at the base of the food chain. It turns out, however, that things were more complicated than that.

Dinosaur extinction

More fancy artwork depicting the K/T apocalypse with a *Tyrannosaurus rex* in the middle of a desert, (an unlikely setting for such a large meat eater) with falling meteorites. Nevertheless, there is the evidence of a big hole in Mexico which shows that a 10–20km- (6–12½ mile-) wide impactor from space did indeed hit the Earth with devastating effect about 65 million years ago at the end of the Cretaceous period and was at least partly responsible for the extinction of a wide range of creatures, including the dinosaurs.

strength of the material of the impactor that, while the front of it decelerates, the back does not, resulting in incredible compressive forces being generated and release of heat energy at several thousand degrees Celsius. The effect would be catastrophic, with the impactor exploding, vaporizing, and releasing kinetic energy (an order of magnitude higher than that of a hand grenade) which is expressed by crater formation. Not only was a crater formed, but also its rim was pushed up to form a wall 8km (5 miles) high within 20 seconds of impact, only to collapse within a couple of minutes.

The initial crater was around 118km (73 miles) in diameter and 35km (21 miles) deep, enough to push the top of the mantle down by about 2,000m (6,500ft). A hole in the Earth's surface of this size cannot be sustained, so that its walls instantly collapse into the crater, generating circular normal faults with displacements of up to 3500m (11,480ft) in one go. These lead to collapse rings which extend to 100km (60 miles) from the epicentre (ie 200km (120 miles) in diameter). A collapse on this scale also refills the crater with wall debris and pushes up the centre of the new crater floor.

The impactor hit a shallow sea, then crashed through the sea floor into limestone sediments below. Shock waves transmitted through ocean waters are thought to have generated tsunamis perhaps up to 1,000m (3,280ft) high, which would have devastated coastal regions all around the Gulf of Mexico and perhaps further afield. The material from the impactor and from the crater was ejected back through the atmosphere; some of it may even have reached the Moon. There was such a mass of material that fall-back velocities have been estimated as high as 8–9km (5–5½ miles) per second. Calculations suggest that the reentering rock material would have behaved like shooting stars, generating white heat energy, two-thirds of which would have been transmitted through the atmosphere towards the Earth's surface. Within 1,000 seconds of impact, there would have been two to three hours of white heat energy bombarding the Earth's surface. Such flash heat radiation would have caused global wildfires, burning off vegetation. The evidence for this is the occurrence of very widespread charcoal lying

The size of the impact crater and its associated collapse rings indicated that the original size estimates for the impactor were pretty accurate. A body this size would have entered the Earth's atmosphere at speeds of more than 25km (15 miles) per second and perhaps as high as 70km (43 miles) per second if it were a comet. The energy levels are so high compared with the

just above the boundary within terrestrial sedimentary sequences. In addition, fossil pollen shows that there was widespread deforestation, especially within North America, and replacement of mature plant communities with ferns, which readily recover from wildfire. Only burrowing organisms or those living more than 2–3m (6½–10 feet) below sea level would have escaped the searing heat. Otherwise, the only refuges would have been where there was heavy cloud.

The atmosphere would certainly have been laden with dust for months, but rain would have washed most of this out within six months. A much more damaging effect was the generation of acid rain from the burning of limestone in the atmosphere, releasing sulphur dioxide that then dissolved in rainwater. It may have taken between 10 and 100 years to return to more neutral acidity in rainfall. Global climates would certainly have been disturbed, perhaps by as much as 15°–20°C (59°–68°F).

All land animals of more than 25kg (55lb) became extinct and, because most of these were reptiles, the

path was cleared for a group of small, "hidden" creatures to come out – they were the mammals. Although much vegetation was destroyed or severely damaged, plant life naturally suffers from regular wildfire and has evolved ways of recovering. For instance, there can be regrowth from undamaged root sytems or buried seeds.

The other interesting and puzzling terrestrial survivors of this extinction event were the birds, who might be thought to be among the most vulnerable in such a catastrophic scenario. However, the best statistical evidence for the biological impact of the event comes from the marine record and the tiny unicells, the forams, which kick-started the whole recent investigation of the nature of the K/T boundary extinction. The forams are very abundant microfossils and can usually be recovered by the thousand from suitable marine sediments. Some of these micro-organisms live in surface waters (as part of the plankton), while others live on the sea bed (as part of the benthos), and it was the former group that was more affected by the extinction.

Bug Creek (far left)
Many K/T boundary strata are marine, but at Bug Creek in Montana, USA there are terrestrial and freshwater deposits containing fossils of dinosaurs and mammals. It was thought at first that perhaps this was one place where some dinosaurs survived the extinction event and lived alongside early Tertiary mammals. However, closer analysis suggests that the older dinosaur fossils were reworked by rivers and mixed with those of younger mammals.

Impact-shocked crystal
Rock debris ejected by the Chicxulub impact event 65 million years ago shows signs of severe physical stress which can be produced by no other terrestrial mechanism. Here a quartz crystal, originally from Chicxulub but found in K/T boundary deposits on the Caribbean island of Haiti, is shot through with thin "shock produced" fault planes from the impact.

Volcanism and climate change

Deccan traps

Coincident with the 65-million-year-old impact event was the outpouring of incredible volumes of basalt lavas over the Deccan region of western central India. Greenhouse gasses released by the huge scale of the volcanism may well have contributed to rapid climate change and enhanced the magnitude of the extinction of life.

Geological mapping of the remoter parts of the world has revealed vast areas covered with basaltic lavas known as plateau basalts. These basalts are not necessarily linked to individual volcanoes, but are the product of large-scale fissure eruptions, and their nature has been something of a problem until recent decades.

Some of them, such as the Miocene age Columbia River basalts, which may have poured out 170,000km³ (40,785 cubic miles) of basalt, and the Tertiary (end Paleocene age) Igneous Province of the northwest part of the British Isles (Brito-Arctic region), have been known about for a long time. Since the advent of plate tectonic theory and their accurate dating, however, a new picture has emerged for these extraordinary events which have punctuated Earth history. It turns out that a number of these basalts coincide with both major and minor extinction events.

Major ones include the 65-million-year-old Deccan traps of India, estimated to have erupted some 2,000,000km³ (479,825 cubic miles) of lava, which coincides with the end Mesozoic extinction. There are also the 251-million-year-old Siberian traps (between 2–3,000,000km³ (479,825–719,738 cubic miles) of lava which coincide with the end Permian extinction. The eastern North America plateau basalts of the New Jersey region, which may have originally had a volume of some 2,000,000km³ (479,825 cubic miles), coincide with the

smaller end Triassic (Norian) extinction event. Inevitably, such coincidences have led to speculation that there might be a connection between extinction events and large-scale basalt outpourings.

Rising plumes of heat through the mantle (also known as mantle hot spots) promote such outpourings of lava. They cause doming through expansion of the crust, which in turn stretches the uppermost surface layer. The heat induces partial melting of basaltic rocks at the top of the mantle, and these fluid melts rise to the surface, especially along tension cracks opened by the stretching forces acting on cool, brittle surface rocks. Many past mantle plumes have been precursors to continental rifting and the opening of new oceans in eruptions such as those at the end Jurassic age Parana/Etendeka flood basalts of South America and southwest Africa, respectively, which are linked to the opening of the South Atlantic and those of the end Palaeocene Brito-Arctic Province, linked to the opening of the North Atlantic.

The effects of this kind of volcanicity which relate to global extinction revolve mainly around their gaseous products entering the atmosphere, leading to climate change. Sulphur dioxide and carbon dioxide are volumetrically the most important gases. Sulphur dioxide is a greenhouse gas, and its initial effect is to cause global warming; however, it soon reacts with water in the atmosphere to produce sulphate aerosols which absorb the Sun's radiation, leading to subsequent cooling and acid rain in the longer term (lasting up to 10 years). Any volcanic ash in the atmosphere also promotes cooling, but so far modelling of such pyroclastic elements cannot be linked to extinction events. The release of carbon dioxide has a much longer-term effect, but its cumulative effects could well be climatically significant.

The end Permian Siberian traps are unusual in that they contain large volumes of volcanic ash deposits precipitated from the atmosphere. Their abundance is thought to have been responsible for climate cooling, which some experts think was severe enough to cause glaciation over the period of eruption, some 600,000 years. This, in turn, may have been responsible for the major fall in sea level at the end of the Permian. There is as yet no supporting geological evidence, however, for an end Permian glaciation. Furthermore, there is evidence from a number of sections through boundary strata that, on the contrary, there was a rise in sea level at the time. The most persistent signal associated with the end Permian extinction is global warming, to which the Siberian volcanic activity, which was the biggest outpouring of plateau basalts known, could have contributed. However, the role of methane has also been invoked recently, and this does not emanate from the Siberian fissure eruptions.

The Deccan outpouring of basalts at the end of the Cretaceous also very interestingly coincides with an extinction event. The lava pile reached a peak thickness of 2.5km (1.5 miles), and its total outpouring of some 2,000,000km³ (479,825 cubic miles) of lava is thought to have happened within one or two million years at most. Estimates of the associated outpouring of carbon dioxide amount to some 500,000,000,000km³ (11,956,379,829 cubic miles). Calculations of global warming associated with these figures, however, suggest that it would only amount to 1–2°C (1.8–3.6°F). By comparison, similarly large outpourings of sulphur dioxide could have led to short-term cooling, which cumulatively could have lasted longer because of the time scale of the total eruption.

Surprisingly, then, it seems that so far it is not clear what exactly the effects of such large-scale volcanic eruptions might have had on life forms and how they would have contributed to the extinction events. For instance, sediments formed in between eruptions are fossiliferous in places, containing remains of freshwater fish and amphibians; however, these show little evidence of any extinction, even though they are right in the middle of the event.

The dinosaur record is even more curious in the Deccan, for there are fossil fragments of dinosaur eggshells which occur above an iridium anomaly which presumably was caused by the impact event. If this is correct, then, right in the middle of the Deccan is evidence that dinosaurs might actually have survived

into the earliest part of Tertiary times. Alternatively, they may have been reworked from older layers.

There is independent evidence for global climate cooling at the end of the Cretaceous period before the impact event, and there are faunal extinctions that precede the impact such as the demise of the rudist bivalves and many benthic forams. The onset of cooling also, however, precedes the eruption of the Deccan traps by four to six million years. The cooling trend continued into the earliest Tertiary time (Paleogene), but was punctuated by a brief reversal with a 500,000-year phase of warming at the end of the Cretaceous. This coincided with a fall and then rise in sea level. The warming phase coincides with the eruption of the Deccan traps, which may indeed have been responsible for it.

Mount St Helens

The 1980 eruption of this volcano, dormant since 1857, in the Cascade Range of Washington State, northwestern USA was another reminder of the Earth's continuing dynamism. On 18 May it erupted, sending up a plume of ash 19km (11.8 miles) high. The eruption killed everything within an area of 180km² (69.5sq miles), and spread ash even further. The eruption blew off about 3km³ (0.7cubic milès) of the volcano, reducing its height from 2,950m (9,678ft) to 2,549m (8,363ft).

Ice ages

The discovery that there were ice ages in the geological past was not entirely unexpected because such events had been predicted by scientists such as James Croll (1821–90), who studied the periodic variations in the Earth's orbit around the Sun. However, there was a long-running problem regarding the kind of signature past ice ages might have left in the stratigraphic rock record and its identification and interpretation.

To begin with, most evidence for glaciation was land-based and had little chance of being preserved in the ancient rock record. The recognition of what seemed to be ancient glacial tillites (with dropstones from floating ice) well beyond polar regions then raised the question of how they got there – until, that is, plate tectonics helped to resolve the problem. In fact, mid- to late Carboniferous glacial deposits (some 310–320 million years old), such as the Dwyka Tillite of South Africa, were part of the early evidence used by Alfred Wegener (1880–1930) to argue that there had been what he called "continental drift". Evidence for

Precambrian snowball Ice Age

American geologists Dan Schrag (left) and Paul Hoffman (right) pose next to a dropstone rock layer (with boulders rafted by icebergs) and its overlying dolostone cap rock (representing relatively warm water deposition) in Namibia, southwest Africa, which marks the abrupt end of a 740-million-year-old snowball event.

contemporaneous glaciation had been found in Africa, South America, India, and Australia which only made sense if the continents were clustered together near the south pole at the time.

The most exciting ice age–related discovery is the "Snowball Earth" hypothesis which has been particularly promoted over the past decade or so by American geologists Joe Kirschvink, Paul Hoffman, and Daniel Schrag. The idea is that, between 700 and 580 million years ago, in the late Precambrian, there were perhaps as many as four major glaciations so extensive that polar ice extended right into low tropical latitudes. There may also have been similar events as long ago as 2300 million years, but none of the Phanerozoic ice ages has been anything like as extensive.

The claim is that the Earth may have frozen over completely. If so, then planet Earth certainly would have looked like a snowball from space, a gleaming white ball with the sunny side reflecting back most of the Sun's light energy. The extensive sea-ice cover would have chilled even the salt waters of the oceans to such an extent that the ocean circulation system would have been shut down and cut off from the atmospheric circulation system. At this time, there was no life on land. The big question is what would have happened to life in the oceans. Even in late Precambrian times, there was abundant marine life, but it was almost entirely microscopic. Could anything have survived under such an ice cover? Initial estimates claimed an ice cover up to a kilometre thick and global temperatures hovering around -50°C (-58°F).

The evidence for such low-latitude glaciation has been mounting since 1949, when it was first suggested by Sir Douglas Mawson (1882–1958) to explain late Precambrian glacial deposits in the Flinders Ranges of South Australia. These are marine deposits with dropstones from sea ice, capped by a layer of dolostone carbonates. It is only over the past few decades, however, that it has really taken off as an idea, especially since the advent of plate tectonics and increasingly reliable latitudinal data for continental positions at the time. From measurements of the magnetization of rocks of the times, polar and latitudinal positions can be calculated. From this data, it appears that most of the continents lay in the tropics, having just broken away from an early supercontinent configuration. Now, at least 16 globally separate sites have been discovered which record late Precambrian glacial sediments, many of which were laid down within 10 degrees of the equator. Significantly, the glacial sediments pass abruptly into marine carbonates, indicating a sudden switch to warmer marine conditions. The carbonates also preserve a number of other unusual sediment and geochemical features (particularly carbon isotope anomalies) which are not found in any other Phanerozoic glacial successions. There is also a reappearance of iron formations which are more characteristic of early Precambrian times when atmospheric and ocean chemistry was different.

The impact of the glaciation on the microbiota of the oceans can be seen in a marked reduction in diversity of the microflora and microscopic eukaryotes. Although it might be argued that a complete snowball scenario would have wiped life out altogether, this is highly unlikely. Almost certainly there would have been ice-free tidal shorelines, brine channels, hot springs around volcanic islands, and open areas in the sea ice. Life is almost irrepressible, and, once a glacial event was over, there would have been enhanced productivity in the oceans, fed by an abundance of dissolved nutrients, leading to a burst in oxygen output. The overall effect of the glaciation perhaps held back the evolution of larger multicellular organisms. These do not appear in significant numbers until the enigmatic, soft-bodied Ediacaran macrofossils of around 575 million years ago.

At one time it was thought that a snowball Earth could be explained if the Earth's axis of rotation had been radically different during much of the Precambrian, with the poles lying where the equator is now. However, this possibility is now discounted largely because it would not explain the abrupt onset and termination of the glacial events. Modelling the succession of events shows that, with most continents clustered in low latitudes, the atmosphere would have been cooled substantially because of the higher reflectivity (known

Dwyka Tillite
Near Kimberley in South Africa, well-developed glacial striations scour a late Carboniferous-age (around 310 million years old) rock surface. The surface and its overlying moraine deposits, known as the Dwyka Tillite represent a glaciation that impacted upon the southern continents of Gondwana (Africa, South America, India, Australia and Antarctica).

as the albedo effect) and low cloud cover. In addition, high rates of weathering of the upland rocks of the continents would have lowered atmospheric carbon dioxide levels and promoted an "ice-house" condition, leading to glaciation. No sooner had the runaway glaciation taken hold than the normal processes, which consume the carbon dioxide in the atmosphere, would have shut down. This shutdown allowed carbon dioxide levels to build up again because volcanoes would still pump the gas out into the atmosphere. As carbon dioxide is a greenhouse gas, capable of trapping heat, the rising levels would cause the atmosphere to warm up again and melt the ice, perhaps within a geologically short time — over hundreds or a few thousand years — resulting in a rapid rise in sea level.

There are, however, experts who claim that some of these glacial sequences show pulses of glacial deposits. They argue that these pulses indicate that there were advances and retreats of the ice, similar to the more familiar Pleistocene glacial succession, and that a complete shutdown of the hydrological cycle is unlikely.

Changing sea levels

Geological evidence for significant changes in ancient sea levels has been mounting over the past 200 years. In Britain and Scandinavia, early 19th-century geologists discovered curious stepped features stretching back from present-day coastlines and sometimes reaching tens of metres above present-day sea level. When investigated in detail, these terraces often displayed beach-type sand deposits, including the shells of shoreline creatures, wave-cut notches, and even sea caves and old sea cliffs. Had the sea level fallen since then or had the land risen? Either way what had caused such changes? Were they local or more widespread and perhaps global?

We now know that the question of past sea level change is really quite complex, with several factors involved. Some changes are just regional, such as those of northwest Europe where the land has rebounded since the melting of the Pleistocene ice sheets, which had literally weighed the land down. This isostatic rebound, as it is called, is now a well-understood phenomenon and still continues today. Over and above this process, however, there were also global changes in sea level during the Pleistocene. During glacial phases, so much water was withdrawn from the oceans, cycled through the atmosphere and precipitated as snow to become locked up as glacial ice (on both land and sea) that global sea levels fell significantly. During warm interglacial climate phases, the ice melted and the water was returned to the oceans, causing sea levels to rise again.

The overall changes amounted to as much as 100m (330ft), which might not seem much compared with the depths of the oceans. However, such changes have made enormous differences to the life of the past. The basic reason for this is the nature of offshore submarine topography. The continents extend beyond their immediate coastlines and are fringed by a shallow submarine shelf called, appropriately, the "continental shelf". This slopes gently away at first, but then comes to an end with a marked steepening of the slope at depths of around 200m (660ft), then rapidly descends into the ocean depths of 2,000m (6,600ft) or more. Strictly speaking, the continents end where this break in slope, known as the "continental margin", occurs. The width of the shelf varies enormously depending largely upon the geological environment and the plate tectonic situation.

Where there are converging plate margins, such as around much of the Pacific basin, the shelves are very narrow because subduction of the ocean floor is occurring offshore and is typically marked by a submarine trench. Any sediment shed from the land into the sea is dumped into the trench and has no opportunity to accumulate. In contrast, where there are passive margins such as around much of the Atlantic basin, shelves are much wider and are largely made up of great wedges of sediment from the land. These accumulate in great piles offshore over many millions of years, especially where great river systems, such as the Mississippi–Missouri into the Gulf of Mexico or the Amazon into the Atlantic, discharge into the ocean with huge deltas.

These continental shelf accumulations of sediment are one of the most important geological features of the Earth for a number of reasons. The shelf surfaces are sites where much of the diversity of marine life thrives, especially where reefs develop. The turnover of such huge biotas results in the burial of very large amounts of decaying organic matter, along with their residual hard parts, which are to become the bulk of the future fossil record. Buried organic debris, especially within or around deltas, can lead to the development of hydrocarbons over the long term, providing the circumstances are right. The sediment fringes themselves represent future sedimentary rocks. Such large accumulations are hard to destroy. When the plate configurations and movements change and passive margins become convergent ones with plate collisions, these sediment piles take the brunt of much of the collision forces and are crumpled into new mountain ridges.

The relative shallowness of continental shelf seas means that a fall of 100m (330ft) or so in sea level will expose a significant proportion of the continental shelves around the world and decimate most of the coral reefs and other life forms which normally occupy these areas. On the positive side, for land-living creatures, any such fall would in the long run provide huge new territories to colonize. Perhaps even more importantly, many islands, such as the British Isles and adjacent continents, would become reconnected by land bridges, such as those between Asia and North America, and Asia and Australia. Reconnections such as this have opened freeways for exchange of life, leading to radical changes in dominance and diversity. For instance, sea level changes during the recent Pleistocene ice ages allowed modern humans to migrate into Australia as long as 50,000 or more years ago, but kept modern humans out of the Americas until some 15,000 years ago.

Many decades of effort have been required for geologists to discover what the geological history of global sea level change has actually been. To prove that events have indeed been global rather than regional, evidence from rock strata has to be accurately matched between continents. Sea level curves have now been established for Phanerozoic times, but their accuracy diminishes for earlier periods. Taking present levels as a standard, it appears that sea levels are now at their lowest for most of Phanerozoic time, except perhaps at the end of the Permian. A number of cycles of rising and falling levels have been recognized, with periodicities of tens of millions of years. Over and above this are longer term changes. Tracing back through Cenozoic times, levels generally rise to a peak of nearly 400m (1,300ft)

Raised beach

Scotland, along with Scandinavia, is still slowly rising out of the sea (known as isostatic rebounding) as a result of the removal of the ice load after the Quaternary ice age, which ended around 12,000 years ago. Here in the Western Isles of Scotland an old sandy bay (now grass-covered machair) and surrounding seacliffs now lie a few metres above present-day sea level.

above present at the end of the Cretaceous. Beyond this, levels gradually fall towards the same levels as today at the beginning of the Mesozoic. Looking into the Paleozoic, levels rise again to an all-time high – perhaps up to 600m (1,970ft) above present sea level – at the end of Ordovician times, coinciding with its ice ages.

These phases of much higher sea levels correspond with extensive flooding of the continents by shallow seas, resulting in high diversities of marine life and the deposition of huge amounts of sediment on the continents. Falling sea levels led to the retreat of the seas from the continents, erosion of continental deposits, falling marine diversity, and extinctions. Indeed, most extinction events coincide with low global sea levels. The most dramatic of these marks the boundary between the Paleozoic and Mesozoic eras (ie, the boundary between the Permian and Triassic periods) – the end Permian extinction event.

The Permo-Triassic extinction event

The biggest disaster for life on Earth occurred around 251 million years ago. This extinction event marked the end of Paleozoic marine life and the rise to dominance of more modern sea creatures. Perhaps some 80 per cent of all marine genera became extinct over a relatively short period of time, perhaps a few hundred thousand years.

Some experts claim that the extinction event happened in two pulses separated by perhaps as much as 10 million years, and it is the latter, bigger pulse that

Extinct life forms

At the family level, diversity was nearly halved from around 400 marine families to 200 as a result of the Permo-Triassic extinction event. Those that suffered most were the Paleozoic corals, which were effectively wiped out, as were the last of the trilobites, the rostroconch bivalves, hyolithid molluscs, and the blastozoan echinoderms, plus many crinoids, brachiopods, bryozoans, gastropods, and arthropods. Many of these animals were filter feeders that lived fairly sedentary lives within particular niches on the seabed. In addition, the Paleozoic sharks, rays, and early bony fishes suffered severe losses from which they did not really recover until Cretaceous times.

is taken to mark the end of the era. Other experts dispute this, however, and say that there is no really convincing statistical evidence for two pulses. What is not disputed is the overall decimation of life.

Evidence from land environments is not nearly as good as that for the seas, but does show changes which had significant long-term effects. There was a great increase of fungal spores found in boundary sediments, which shows that terrestrial ecosystems were severely hit. The tree-sized (arborescent) lycopsids were drastically reduced in diversity and eventually replaced by the cycads and conifers. Land-living vertebrates also suffered, especially the Paleozoic amphibians. These latter were tetrapods and much more like reptilian crocodiles than the frog-shaped amphibians we are familiar with today. Indeed, the amphibians have never really recovered their early diversity. The tetrapod reptiles were also hit, especially the cynodonts and gorgonopsids, but new archosaur groups appeared and diversity recovered by early Jurassic times.

So what caused this devastation of life? Surprisingly enough, the question cannot yet be answered simply, and it is one of the outstanding puzzles in the history of life. Unlike the next biggest extinction event at the end of the Cretaceous, we cannot point to an impact event, although many scientists have devoted a large amount of effort looking for evidence of one, which would certainly provide a relatively simple solution. But there is no geochemical signature or impact debris that can be linked with an extraterrestrial event, even though there is significant change in the organic chemistry of the sediments, linked to changes in the ocean enviroment (known as carbon isotope excursion). No large-impact crater of the right age has been found. It must be remembered, however, that if a large body from space had hit the ocean rather than the land, we would not find the crater because the ocean floor of this age has all been subducted. Despite this, there should still be an iridium spike like the one found at the K/T boundary.

What we do find is that there was a vast outpouring of flood basalts in Siberia around this time, and global sea level fell to an all-time low for Phanerozoic time. This coincides with the formation of the supercontinent of Pangea, which straddled the Earth virtually from pole to pole, but had the bulk of the continental landmasses

Continental drift

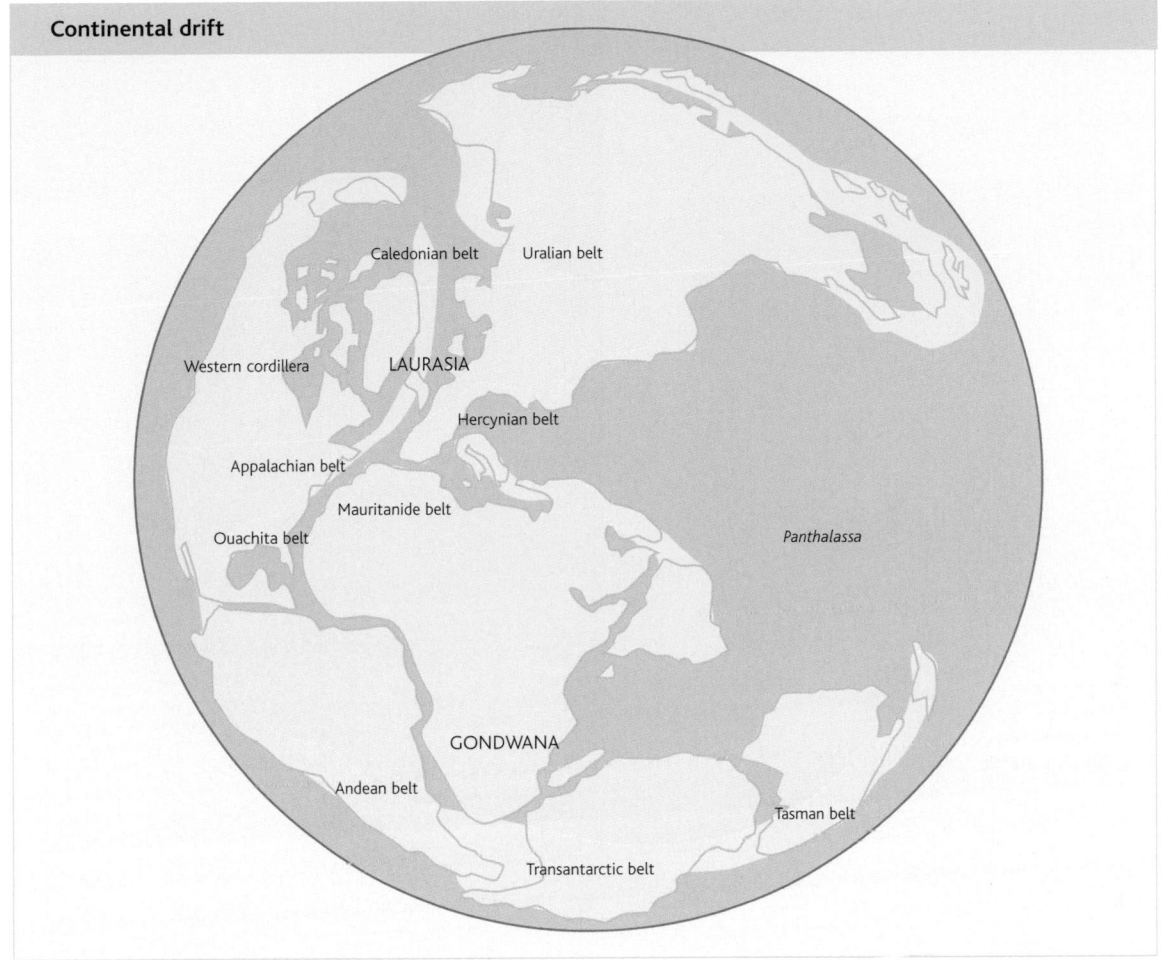

Caledonian belt Uralian belt

Western cordillera LAURASIA

Hercynian belt

Appalachian belt

Mauritanide belt

Ouachita belt *Panthalassa*

GONDWANA

Andean belt

Tasman belt

Transantarctic belt

Pangea

A reconstruction of the supercontinent of Pangea which was assembled some 290 million years ago. It stretched from pole to pole and was surrounded by a single ocean called Panthalassa. Newer mountain belts (eg. Cordilleran-Andean) and older ones (Caledonian-Appalachian) are shaded.

concentrated in high latitudes within the southern hemisphere. The effect of assembling such a global jigsaw of continental plates is to reduce drastically the total length of coastlines and the area of continental shelves. Consequently, habitats for the bulk of marine life capable of being fossilized were radically reduced. On land, the formation of such a supercontinental mass has far-reaching effects on global climates and leads to the establishment of very dry interior landscapes – deserts. In addition, the outpouring of huge volumes of greenhouse gases such as carbon dioxide from the Siberian lavas would have promoted global warming and aided desertification,

Evidence from marine sediments, especially a carbon isotope shift at the boundary, has been linked to a release of methane gas and suggests that the oceans and perhaps the shallow seas of the time became anoxic (lacking in oxygen), causing the reduction in marine

diversity. This raises the question of the origin of the methane. A volcanic source has been ruled out, but the recent discovery of methane hydrates buried deep within sea-floor sediments – typically in water depths of 300m (980ft) or more – raises an intriguing new possibility. Theoretically, a rapid fall in sea level could lead to a sudden release of enormous volumes of methane into the ocean and then into the atmosphere. As a greenhouse gas, methane is 3.7 times more powerful in its warming effect than carbon dioxide. Consequently, not only would life in the oceans have been catastrophically damaged by the methane release, but so, too, would terrestrial life have suffered similar devastation, as a result of rapid warming of global climates. It is not yet clear, however, how much of the methane would have reached the atmosphere and how much would have been oxidized to carbon dioxide in the oceans.

6

Life's early age

The Paleozoic era of ancient life was not investigated in any detail
until the 19th century, although some strata were familiar prior to
that through their economic use. For instance, outcrops of coal
associated with strata of Carboniferous age were exploited in Europe
from the late 18th century onwards. Their excavation soon brought
numerous fossils to light, especially those of the plants that made up
the coal deposits. However, much of lower Paleozoic-age strata
formed mountainous uplands in northwestern Europe, having been
subjected to folding, faulting, and metamorphism. Consequently, their
complicated structure was not subdivided into geological systems and
periods until the latter 19th century.

As this mapping progressed and the Cambrian, Silurian, and
eventually Ordovician periods were defined, their fossil faunas were
discovered and described. Practically all the fossils were new to
science, as they belonged to long extinct taxa.

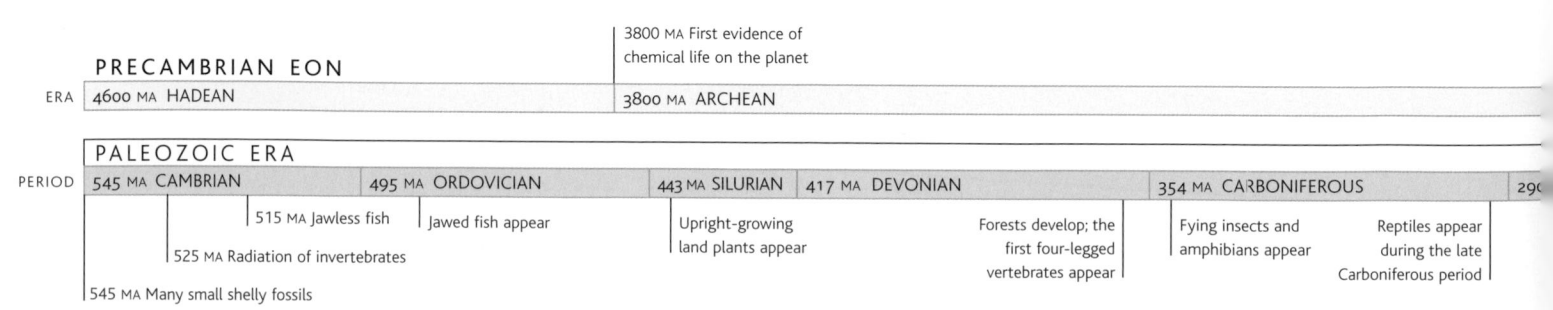

PRECAMBRIAN EON

ERA	4600 MA HADEAN	3800 MA ARCHEAN

3800 MA First evidence of chemical life on the planet

PALEOZOIC ERA

PERIOD	545 MA CAMBRIAN	495 MA ORDOVICIAN	443 MA SILURIAN	417 MA DEVONIAN	354 MA CARBONIFEROUS	29C

515 MA Jawless fish

Jawed fish appear

525 MA Radiation of invertebrates

Upright-growing land plants appear

Forests develop; the first four-legged vertebrates appear

Fying insects and amphibians appear

Reptiles appear during the late Carboniferous period

545 MA Many small shelly fossils

Lifetimes were spent investigating the amazing lower Paleozoic fossils in regions such as Bohemia in central Europe, (described by Joachim Barrande) and New York State (described by James Hall). It soon became evident that while landgoing vertebrates were found in strata as old as the Carboniferous, they were not present in lower Paleozoic strata, although fossil fish were found. Their record extended back into the Silurian, with many strange armoured creatures without bony jaws or teeth. Mostly, the lower Paleozoic strata were filled with marine fossils; even primitive land plants did not seem to be present until Devonian times. Clearly, some sort of progression in ancient life was gradually emerging, although it was too vague to be of use to Charles Darwin as he marshalled evidence for his theory of evolution. In fact, Darwin went to great pains to avoid referring to the fossil record because he knew it was still too full of gaps and problems. Today fossils provide countless examples which reinforce the Darwin/Wallace arguments for the origin of species.

Fossilized fern

Many fossil plants, such as this *Alethopteris* frond found in late Carboniferous coal-bearing strata from Pennsylvania, USA look superficially like modern ferns. However, they are gymnosperms (seedplants) and some, like *Alethopteris*, whose fronds grew to 7m (23ft) long, were tree-sized.

1200 MA The first multi-celled organisms date from the middle of the Proterozoic period

610 MA The first large marine animals appear

PHANEROZOIC EON

2500 MA PROTEROZOIC

545 MA PALEOZOIC | 248 MA MESOZOIC | 65 MA CENOZOIC

TODAY

MESOZOIC ERA

CENOZOIC ERA

MIAN | 248 MA TRIASSIC | 205 MA JURASSIC | 142 MA CRETACEOUS | 65 MA PALEOGENE | 23.8 MA NEOGENE

The first dinosaurs and mammals appear

Birds and flowering plants appear in the late Jurassic period

124 MA First placental mammals

Primates and songbirds appear

5 MA First hominids appear on earth

Major extinction event

The discovery of deep time

Cambrian heartland
The Harlech Dome in North Wales formed part of Adam Sedgwick's Cambrian System, a period of geological time which contains the first abundant shelled fossils.

James Hutton (1726–97) was one of the first earth scientists to peer into the "abysmal depths" of the geological past and declare that he "could see no vestige of a beginning, no prospect of an end". Hutton had a brilliant scientific imagination and the ability to grasp the implications of different configurations and

relationships to be seen in rock strata. He realized that inordinate amounts of time relative to the human or even historical scale are required to produce the rocks that can be seen at the Earth's surface. By the early decades of the 19th century, the youngest strata, widely exposed throughout the Mediterranean and northwest European regions, had already been explored and subdivided into a hierarchy of sequences which could be matched between different rock outcrops, across countries, and even between countries and continents. The history of the Earth, like that of humankind, was divided into recognizable units – systems of rock strata that represented periods of geological time.

By the 1820s, geological exploration and mapping by the likes of Cuvier, Brongniart, and Smith had outlined the distribution in time and space of strata belonging to what we now recognize as the Cenozoic and Mesozoic eras. It was also recognized that there were still plenty of strata lying below the oldest Mesozoic rocks. Since Werner's day, these had been called the "Transition", or "Grauwacke", series, and they were known to contain some organic remains. These strata in turn lay upon Primitive rocks which include gneisses, schists, slates, basalts, and, oldest of all according to Werner, granite.

Smith's geological map of Britain showed that much of Wales, the Lake District, and southern Scotland was made up of Transition rocks which he had not subdivided. A substantial part of northern Scotland was made of a complex mixture of Primary rocks which John MacCulloch (1773–1835) was investigating. Clearly the *terra incognita* that needed to be explored first was that of the Transition series rocks.

One of the centres of the geological world of the early and middle decades of the 19th century was the Geological Society of London. Founded in 1807, the society was a breakaway from the Royal Society, and most of the active geologists in Britain were members, including eminent figures such as Charles Lyell, William Buckland, Richard Owen, John Phillips, Thomas Henry Huxley, and Charles Darwin. The most famous geologist not to be elected a fellow of this middle-class club was William Smith himself, who, as a practising engineer and surveyor from humble origins, was beyond the social pale – that is, until he was belatedly recognized by the society in 1831.

Among the young geological "turks" of the 1820s were Adam Sedgwick (1785–1873) and Roderick Murchison (1792–1871), who were determined to make their mark in this new science. Although from very different backgrounds, the two became active collaborators and friends. Sedgwick was from rural Cumbria, where his father was a clergyman and schoolmaster, and as a clever student Sedgwick won a scholarship to the University of Cambridge, where he studied theology and mathematics. When the Woodwardian Professorship of Mineralogy became vacant in 1818, Sedgwick applied and was appointed, even though he had not formally studied geology. Like most Cambridge academics of the time, he took holy orders and had to balance his clerical duties with teaching and lecturing for his college. Sedgwick took his duties seriously, however, and was very capable, and as a consequence he soon studied what was known of geology and found the profound unknowns of the new science challenging.

By contrast, Murchison came from the landed gentry of Scotland; however, being a younger son, he had to make his own way and served as a cadet officer in the Peninsular War (1807). On making a good marriage, he not only acquired a very accomplished wife, but also enough money to be independent. His wife encouraged him to do something useful rather than indulge in the usual round of hunting, shooting, and fishing which most men of his class practised. Murchison attended Geological Society meetings in 1824 and soon became acquainted with the leading lights, including Sedgwick. He toured the Scottish Highlands with Sedgwick in 1827 and parts of France and Italy with Lyell in 1828, and generally educated himself in the rock business.

By 1830, Sedgwick and Murchison had decided to attack the Transition series in Wales. The novice Murchison was to work down the stratigraphic sequence from the known part of the rock succession in the border country between England and Wales into older unknown strata. The more experienced Sedgwick was to

tackle the structurally complex older rocks in northwest Wales and work up the succession with the intention of joining up with Murchison and hopefully identifying any natural subdivisions between sequences of strata along the way.

Murchison had the easier task and was able to make rapid progress – the strata were full of fossils and had relatively simple structure. He was astute at making useful social and scientific connections with local experts, of which there were many in the Borderlands, where country livings gave local clergymen plenty of

Cnemidopyge
This 3cm- (1¼in-) long fossil arthropod (from Welsh Ordovician strata), is typical of the extinct sea-dwelling trilobites with a three-part exoskeleton – comprising a head shield (here with an anterior spine), a body made of hinged plates that allowed it to roll, and a tail piece that protected the underside of the head when it rolled up.

Welsh Silurian strata
Exposed on the Welsh
Atlantic coast, these
tectonically tilted sandstones
and shales contain fossils
such as the extinct
graptolites, which show them
to have been early Silurian
sea-bed deposits, now known
to date back some 435
million years.

time to investigate God's handiwork. Many of them sincerely felt that it was their duty to uncover the "testament of the rocks" and reveal the details of the formation of the Earth only briefly alluded to in the Genesis account.

Sedgwick had a much more difficult time, as the geological structures he was facing were often complex and fossils rare, so that it was difficult for him to match sequences of strata from one place to another. His was truly pioneering work, and he made slow progress, especially when it came to describing the fossils which were so critical for identifying and correlating the strata.

Nevertheless, by 1834 the two young geologists had met up and concluded that they could identify two distinct and sequential systems of rock strata. Murchison's was the younger, and in 1835 he named it the Silurian rock system after a Romano-British hill tribe, the Silures. Sedgwick's older Cambrian rock system was also named in the same year, after the Roman name for Wales. Murchison noted that the top of "his" Silurian system passed up into the strata of the Old Red Sandstone and that its base passed down into sandstones and shales belonging to Sedgwick's Cambrian. Furthermore, there were still more rocks below the base of the Cambrian, and they became generally referred to as "Precambrian", to replace the old term Primitive.

By 1839, Murchison published a large book entitled *The Silurian System*, in which he described the various subdivisions of the strata and illustrated their features along with the all-important fossils contained within them. In a series of technical papers, Sedgwick described the details of his strata, but omitted to describe and illustrate the fossils. The two friends went on to see how the Old Red Sandstones fitted into the scheme and expanded their work into Devon in southwest England, where these strata are well exposed. By 1839, Murchison and Sedgwick jointly proposed that the succession of slates, sandstones, and limestones in Devon which lay stratigraphically above Silurian strata could be distinguished as another younger system in its own right. The two travelled to the Continent and found similar rocks and fossils in the Rhine and Eifel regions of Germany. They named the new system the Devonian, after the English county, and showed that its

strata passed up into the younger Carboniferous rock system which had been named in 1822 by two other British geologists, the Reverend W. D. Conybeare and W. Phillips. Thus Murchison and Sedgwick could truly claim that the old term "Transition series" was no longer needed.

The partnership was not to last much longer because of a growing dispute over the boundary between the Cambrian and Silurian systems. Murchison expanded the Silurian downwards, claiming that fossils that he had described were to be found in the upper part of the Cambrian, therefore making it Silurian, and that furthermore the origin of life was to be found within the Silurian. Even when some of Sedgwick's supporters described fossils from the Cambrian strata in the 1850s, Murchison tried to claim that they were Silurian in age. When Murchison became director of the British Geological Survey, he was able to ensure that his version was imposed upon all official geological maps. The growing acrimony between the two men led to their complete estrangement, which was only partially resolved in their old age. Murchison, meanwhile, had travelled to Russia and carved out another system of strata above the Carboniferous, which he named the Permian after the city of Perm in the Caucasus.

Creation of the Ordovician

In the 1870s, Charles Lapworth (1842–1920), a Scottish schoolmaster turned geologist, finally resolved the boundary dispute between the Cambrian and Silurian by carving out the Ordovician system between the two. Lapworth had made detailed studies of an extinct group of fossils known as graptolites, which are often abundant in these ancient strata. Combined with growing evidence of fossils being widely distributed throughout Cambrian strata – especially in Bohemia (part of today's Czech Republic) and New York State – Lapworth claimed that three divisions based on the succession of fossils could be recognized. These were the Cambrian (then thought to contain the oldest fossils), Ordovician, and Silurian, which together comprised the Lower Paleozoic. This is now the internationally accepted arrangement. The Burgess Shale of British Columbia (a World Heritage Site) and the Chengjiang strata of China (which has produced some of the most famous fossils we have) are both acknowledged to be part of the Cambrian system, which represents a period of time from around 545–495 million years ago.

Fossil fuels fire the Industrial Revolution

The Industrial Revolution was integral to the building of our modern world, with all its benefits of technology, science, and medicine, and the drawbacks of social upheaval, industrial poverty, and environmental damage. The earliest phase in western Europe began in the late 18th century, continued into the 19th century, and was largely based on geological factors and the ready availability of coal and iron ore in conjunction with navigable waterways, nearby ports, and markets.

All these factors were found in Shropshire, England, where local names such as Ironbridge and Coalbrookdale recall the start of the Industrial Revolution. Local deposits of coal and iron ore led to the establishment of foundries, the cast-iron products of which were carried by river down to the port of Bristol, which exchanged its infamous and fading trade in slaves for the rising trade in iron goods. The availability of coal was fundamental to the success of the enterprise, and the growing need for ever greater supplies fed a frenzy of exploration to find new sources of the "black gold".

Across northwestern Europe, the growth of the Industrial Revolution was based on this availability of coal, which was mostly Carboniferous in age. Coal-bearing strata were found scattered across northern France, Belgium, Germany, central and northern England, South Wales, and central Scotland. Even when the Industrial Revolution hit North America, the early supplies of coal in the east of the continent, scattered from Nova Scotia down to Pennsylvania, were Carboniferous in age (the system is subdivided into the Mississippian and Pennsylvanian in the United States). The problem was that rocks of this age (now known to be 354–290 million years old) only have a limited exposure at the surface; much more of it is hidden and deeply buried beneath younger strata. The question was how to find these valuable buried coal-bearing strata and then how to extract the coal. Huge sums of money were wasted as landowners frantically dug pits and shafts in the hope of finding coal on their land when the simplest of geological surveys would have shown the

search to be pointless. Inevitably, there was a demand for proper geological surveys and mapping, which initially were carried out by surveyors such as William Smith, until state-run geological surveys were established (in 1835 for Britain).

Academic interest in coal and coal-bearing strata was much older. Edward Lhwyd (1660–1709), the distinguished naturalist and keeper of the Ashmolean Museum in Oxford described and illustrated fossil plants and insects from Coal Measure strata in 1699. He thought, however, that they had grown within the rocks from seed derived from the living ferns they resembled, seed that had been washed through crevices into the rocks. By 1804, the German naturalist Ernst von Schlotheim (1764–1832) had shown that the Coal Measure plants of Thuringia in central Germany had a genuine resemblance to the living tree ferns of tropical regions, but that in detail they were different and represented a totally extinct ancient flora.

By 1828, the French botanist Adolphe Brongniart (1801–76) had distinguished four separate phases in the development of plant life, beginning with those primitive fern-like ones of the Upper Paleozoic, followed by the first appearance of conifers, then cycads in the Mesozoic, and finally the flowering plants in the Cenozoic. Brongniart argued that the profusion of giant tree ferns, clubmosses, and horsetails in the Coal Measures indicated that the climate of the time must have been as hot as the tropics are today. The implication was that, as such deposits were found in quite high latitudes, the Earth must have been much hotter in the remote past. His was one of the first analyses of ancient climate, and he was right to connect the Coal Measure plants with tropical climates. It was not until the acceptance of the plate tectonic theory, however, that it was finally realized that climates had not changed that much in the past, but rather the continents themselves had moved.

The growing exploitation of Carboniferous coal gave a great boost to the study of the fossil plants from

which coal is made, but associated with the plants were some intriguing animal fossils. In 1852, Charles Lyell visited North America, landing at Halifax, Nova Scotia, where he was met by John William Dawson (1820–99), a superintendent of education who took a keen interest in geology (he later became professor of geology at McGill University). Together they visited the excellent coastal exposures of Carboniferous coal-bearing strata at Joggins, which are several thousand feet thick. Fossil tree stumps, up to 8m (26ft) high, could still be seen in their original position of growth. The fossil roots were known

as *Stigmaria* and could be seen embedded in clayey soils with the trunks growing up through coal seams. These were in turn interbedded with shales and sandstones. The distinctively patterned trunks were generally found separated from the roots and had been given a different name, *Sigillaria* – although, in 1846, the botanist Edward Binney (1812–81) demonstrated that they were related to the same plant. Nevertheless, here was unequivocal proof of the association. Also, Lyell and Dawson noted that the interior of the trunks was filled with sediment with just an outer cylinder of the "bark" converted to

The Black Country

Extensive exploitation of coal in the Industrial Revolution soon changed rural landscapes into heavily polluted "black country" in the manufacturing regions of Europe and North America in the 19th century. The rest of the world soon followed. Within a hundred years we have depleted coal which took millions of years to form.

Coal detail

A section of coal, seen under a microscope at high magnification, shows part of a plant (orange in colour) which has been flattened and buried in a matrix of organic plant debris. The details of coal structure were first worked out by the paleobotanist Marie Stopes, (1880–1958) who is perhaps better known as a pioneer of birth control.

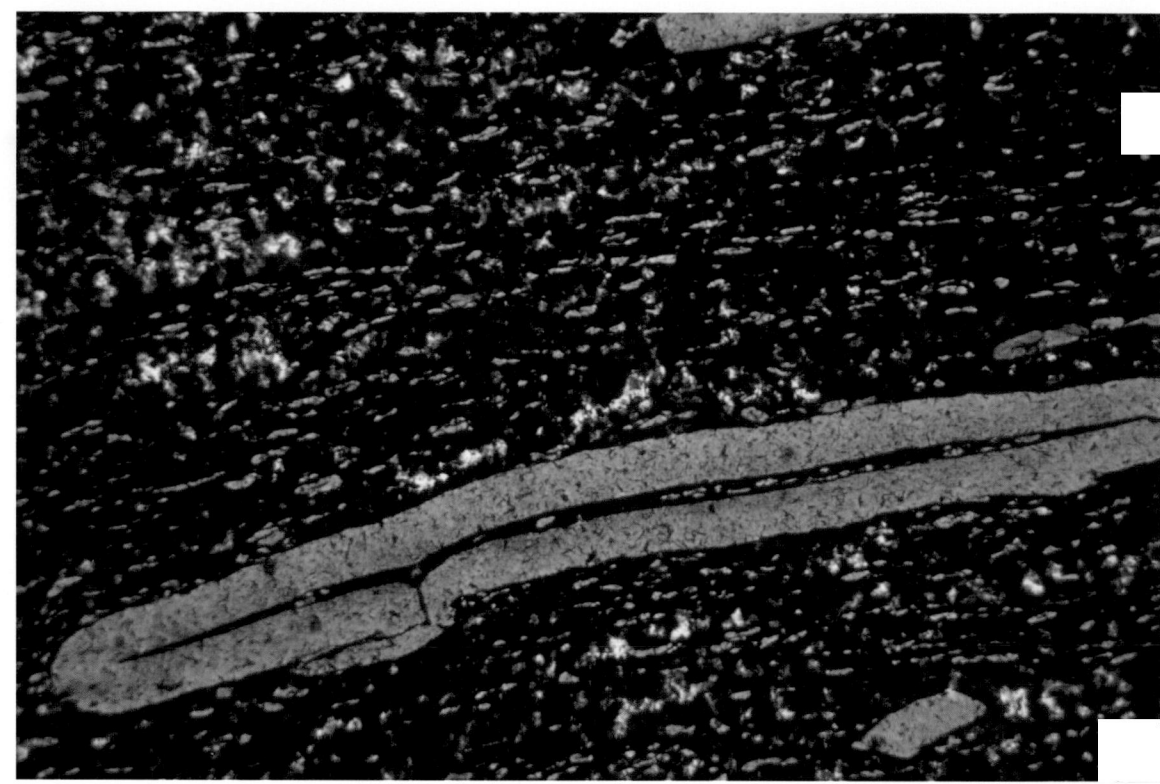

coal. Clearly, when the trees had died in the swamps in which they grew, the interior had rotted away and later been filled with sediment.

Thinking that other kinds of fossils might be preserved within the tree's sediment infill, Dawson and Lyell dug several trees out of the cliffs. On breaking open the sandstone infill, they found fragments of fossil fern fronds (*Sigillaria* and *Calamites*), bits of charcoal, and some bones which even they could see were not fish bones as they expected. They then found some jaws and teeth that clearly belonged to a small tetrapod, which they assumed to be an amphibian such as *Labyrinthodon*. Lyell became very excited by the find. As Dawson recalled, "His thoughts ran rapidly over all the strange circumstances of the burial of the animal, its geological age, and its possible relations to reptiles and other animals, and he enlarged enthusiastically on these points." Noting the surprise of the local resident who was assisting them, "he turned to me and whispered, 'The man will think us mad if I run on in this way.'"

Lyell and Dawson described their new tetrapod as *Dendrerpeton*, thinking that it was a reptile, but it was later shown to be an amphibian. Nevertheless, Dawson

revisited the locality in 1859 and found another small tetrapod about 20cm (8in) long, which he called *Hylonomus* and which was then the oldest known reptile. One of their possible interpretations of the find was that the animals had fallen into the hollow trunks and had not been able to escape; however, recent research has revealed a more dramatic story. The presence of the charcoal shows that wildfire was common in these early tropical forests, sparked off by lightning strikes during storms. The animals almost certainly used the tree trunks for shelter from predators and other adverse conditions, but were killed by a forest fire while hiding there.

The fossil charcoal from such sites is botanically of great interest, as it can preserve remarkable detail of the anatomy, down to the cellular level, of the plant tissue from which it is made. Also, recent investigation of the relative density of stomata – the leaf pores plants use to regulate their "breathing" (gas exchange) – provides a proxy measure of atmospheric conditions and the climates of the time. Basically, when carbon dioxide levels are high, the plants produce fewer stomata, and vice versa.

In late Devonian times, the tree-sized plants of the first forests had relatively high numbers of stomata because carbon dioxide levels were high and climates were globally warm. During the early part of the Carboniferous, carbon dioxide levels started to fall dramatically and climates cooled, switching from a greenhouse to an icehouse state. Sea levels were high, and vast areas of the continents were flooded with shallow seas in which corals flourished and limestones were deposited. Climates became even cooler, more so than those of the present day, and sea levels began falling, creating vast areas of swamps in low-lying continental interiors and around the receding coasts. In tropical regions, these were the environments in which the coal measures developed.

The density of stomata on the leaves of Coal Measure plants becomes much higher, showing that atmospheric carbon dioxide levels were very low and oxygen levels were high (about 35 per cent compared with 22 per cent today). Sea levels frequently fluctuated as they fell, so in coastal regions there were constant changes between shallow marine and shallow swampy conditions, which built up the vast thicknesses of coal measures with their alternations of clays, coals, shales, sandstones, and occasional limestones. The deepening "icehouse" state set off glaciation in high latitudes that persisted into Permian times, although in tropical regions the coal measure forests also continued to flourish. Some of China's vast coal deposits, (producing over 1,000 million tonnes per year), are of Permian age. However, the end Permian extinction saw the biggest change in global vegetation with the loss of the tropical giant clubmosses and high-latitude glossopterids. Of 19 seedplant families (gymnosperms), only three survive into the Triassic – there are almost no early Triassic coal deposits.

Around Edinburgh, the capital of Scotland, limestones of lower Carboniferous (Mississippian) age record some of the critical changes in the life forms that were evolving at this time. Here, the limestones are unusual because they are freshwater and were deposited in lakes around the active volcanoes of the region. The mineral-rich soils promoted a lush vegetation of clubmosses (lycopsids), horsetails (sphenopsids), ferns (pteridosperms) etc. The rotting leaf litter provided home and food for early arthropods such as mites, millipedes, scorpions, and the extinct eurypterids (up to 3m (10ft) long). Crustaceans, shellfish, and fish were abundant in the lakes and rivers. Most interesting perhaps has been the discovery by Stan Wood, a Scots professional collector, of several early skeletons belonging to some amphibious tetrapods.

Varying between about 20cm (8in) and 50cm (20in) long, these little salamander-like animals are very important to our understanding of the evolution of the first egg-laying (true amniote) reptiles. Detailed examination has shown that *Balanerpeton* is clearly an amphibian, as is *Westlothiana*, which was at first thought to be the most primitive reptile ever found; however, *Silvanerpeton* and *Eldeceeon* are both anthracosaurs, as is *Hylonomus* from Joggins in Nova Scotia. Dated at around 338 million years old, these Scottish anthracosaurs are some of the earliest and most primitive reptiles known. Technically they are referred to as reptiliomorphs because their true amniote condition cannot be proved, but they do show significant advances on the amphibian structure.

Tree stumps

Coastal cliffs at Joggins, Nova Scotia, Canada expose Carboniferous-age strata with coal seams and the fossil casts of trees still in their life position. When Charles Lyell and John William Dawson examined some of these fossil trees in the 1850s they found the fossil skeletons of small lizard-like vertebrates inside some of the hollow stumps.

Climate change and shifting continents

Transatlantic links

Satellite images of north-eastern North America and western Europe spliced back together roughly as they were before the North Atlantic opened 90 million years ago. The 400-million-year-old lower-Paleozoic mountain belt ran from northern Norway down the Scandinavian peninsula through Scotland and northwest Ireland, into Newfoundland, and on into the Appalachians in North America.

The most detailed early investigation and mapping of Paleozoic strata were carried out in western Europe in the early decades of the 19th century and soon after in northeastern North America. From the geological study of the rock succession, geologists tried to make sense of their environments of deposition and the extinct life forms which inhabited the lands and seas of the time. The deeper and more ancient the strata, however, the more puzzling the story. The division of the stratigraphic succession into systems representing long periods of geological time was facilitated by marked differences between groups of strata.

At the top of the Paleozoic lay strata which Murchison named as belonging to the Permian system. In western Eurasia, the youngest strata are dolomitic limestones with economically valuable deposits of evaporite minerals such as gypsum and anhydrite. They had been deposited in warm, shallow seas with high evaporation rates. Below lay vast thicknesses of red-coloured sandstones which were clearly terrestrial and were perhaps desert deposits. The relatively rare fossils, including curious large terrestrial tetrapod reptiles, seemed to confirm this. Below the Permian lay the older Carboniferous system with its valuable terrestrial coal

deposits, interbedded with shales and sandstones in the upper part (Pennsylvanian). Again, interpretation suggested that the original vast coal measure forests must have been equivalent to today's tropical rainforests. And yet coal-bearing strata were found in Scotland, Nova Scotia, and even much farther north in Svalbard (around 77° north), well within the Arctic Circle.

The lower part of the Carboniferous (Mississippian) is largely made up of limestones and shales, deposited in shallow seas and often containing extensive highly fossiliferous reefs indicative of warm waters. Yet deeper and older lay the rocks of the Devonian system, which had been the subject of considerable argument in Britain. It turned out that the Devonian strata of Wales and northern Britain were somewhat different from those of Devon in the southwest of the country, and yet they were contemporary. The north had been land with substantial rivers and lakes depositing red sandstones (originally known as the Old Red Sandstone) which entombed primitive fish and occasionally plant remains. By contrast, the contemporary strata of southwest England were largely marine with limestones and shales.

Below the Devonian lay the predominantly marine Silurian and Cambrian strata, between which the Ordovician system was interposed by Charles Lapworth to resolve the dispute concerning the Cambro–Silurian boundary. Like most of the Cambrian and Silurian strata, the Ordovician deposits are marine, but include considerable thicknesses of volcanic rocks which were terrestrial lavas and ashes. Altogether, these latter lower Paleozoic rocks do not provide many clues about the climates in which they were deposited, but they do provide some information. Many mid-Silurian strata across North America and Europe contain coral reef limestones, indicative of deposition in warm subtropical waters. However, older Silurian and Ordovician sedimentary rocks across the same area lack reef limestones and contain less diverse faunas. Indeed, it has long been recognized that there was some kind of extinction event at the Ordovician/Silurian boundary. In recent decades the discovery of glacial deposits in North Africa show that there was a late Ordovician ice age.

For many years these regional successions of Paleozoic strata, such as in North America and Europe,

were considered indicative of changes in global climates. For instance the British sequence seemed to show sequential changes from cool or cold climates in late Ordovician times becoming warmer through the Silurian, into hot and semi-arid in the Devonian, followed by hot and wet in the Carboniferous, before repeating hot and dry in the Permian and Triassic. However, as we shall see, the story is more complicated and more interesting.

Many of the rocks of the western European Paleozoic are folded and faulted, often intensely, along a general northeast to southwest trend. They stretch in a belt from northern Scandinavia down through Scotland, northern England, northern Wales, and Ireland. On the other side of the Atlantic in North America, rocks of similar age and general kind also form an ancient mountain belt stretching from Newfoundland, New Brunswick, and New England down into the Appalachians.

It was originally thought that such mountain belts originated as large, elongate marine basins (called geosynclines), into which huge quantities of sediments were poured. From time to time, the basins were infilled, explaining how deepwater strata passed up into shallow water deposits; however, then the weight of accumulated sediment further depressed the basins, allowing a return to deeper water and more sediment infill. Finally, it was imagined that the basin sediments were compressed by lateral forces and folded and faulted. There were numerous problems with this theory. For instance, in the case of the British sequence, there was good evidence that many of the sediments were derived from the west and northwest, but today there is no landmass from which they could have been derived. Similarly, the confining forces required pressures coming from the northwest where the Atlantic lies. Only in recent decades has it finally become clear how these puzzles, the changes in climate and links between the geology of northwest Europe and northeast North America, could be explained.

When the Roman Emperor Hadrian ordered a wall to be built across northern Britain around 125AD, little did he realize its deeper geological significance in marking a fundamental divide in structure and geological history. Scotland and northwestern Ireland have a very special relationship with North America. They are actually a rocky splinter of the great North American continent (called Laurentia), which broke away when the North Atlantic opened up around 95 million years ago. England and Wales are even more exotic and originally were attached to North Africa and the ancient supercontinent of Gondwana around 460 million years ago. Both Gondwana and Laurentia lay south of the equator and were separated by an ocean called Iapetus.

From a study of rocks on both sides of the Atlantic and evidence provided by rock magnetism, which records ancient pole positions, geologists have worked out that the present arrangement of Britain and Ireland was stitched together around 400 million years ago. From late Cambrian and early Ordovician times, when they were close to the South Pole, England, Wales, and bits of western Europe (together known as Avalonia) broke away from the great southern supercontinent of Gondwana and swept northwards. Their movement was like a windscreen wiper, sweeping across, closing the Iapetus Ocean by subduction of its ocean floor rocks and eventually crashing into North America in mid-Silurian times. At last, the missing northwest opposing force for the folding had been found: it was Laurentia. Around the same time, another plate called Baltica, comprised of today's northern European Baltic countries, also moved towards Laurentia and eventually collided with it.

The process of subduction generated vast quantities of heat energy at depth, partially melting some rocks. The hot molten rock (magma) rose through the crust and broke through Avalonia as a series of island arc volcanoes in Ordovician times. The collision threw up a mountain belt the remnants of which persist as part of our common geological heritage, the heartlands of the Romantic poets – the Scandinavian mountains, the Scottish Highlands, the Lake District in the north of England, and Snowdonia in Wales. The 19th-century poets and painters of the Hudson River school in the United States celebrated the North American continuation of the range. If you join the continents of North America and Eurasia together by closing the Atlantic, it is immediately obvious how this Caledonian mountain belt continues from the British Isles and Ireland into Newfoundland and on into the Appalachians.

The formation of this new landmass and mountain belt in Devonian times provides a source and explains how sediment could be derived from the northwest to fill the great rivers and lakes of northern Britain and Scandinavia. By this time, Laurentia had moved north and was entering the tropics. It continued across the equator in Carboniferous and Permian times, explaining how the coal measures and desert sandstones were formed. There did not have to be radical global climate change because the continental plates have themselves moved through thousands of kilometres over geological time.

Geologists have struggled for years to pin down the present location of the original suture line between North America and Avalonia; the fact that it is deeply buried beneath younger rocks makes their job more difficult. New geological research on the rocks and structure of the Isle of Man, which lies between Wales and Ireland, has shown that the join lies very close to the north of the island and passes through the Solway Firth and across the Border country to Berwick on Tweed. By strange geological accident, this deeply hidden geological boundary is remarkably close to our present "artificial" historical boundary and not far from Hadrian's Wall.

The formation of the Scottish side of the suture, the Southern Uplands – the rolling hills so beloved by Walter Scott and Robert Burns – has also been recently reassessed. These hills are made up of a thickness of more than 6km (3¾ miles) of Iapetus ocean-floor muds, originally laid down between 470 and 430 million years ago. The sediments were scraped off the sea bed as the ocean floor disappeared through subduction below North America. Vast fault slices of sediment were slammed together to form a great wedge across the width of the Southern Uplands into counties Down and Monaghan in Ireland. The unassuming and picturesque landscapes of the Borderlands, with their rolling green hills, belie the underlying collision of continents that formed them. Their unruly geological heritage seems to have persisted through the ages, however, with the Border country being fought over for some 2,000 years.

Ironically, as we have seen, the universally recognized geological periods of time when all this happened, the Ordovician and Silurian, were named after warring Romano–Celtic hill tribes the Ordovices and Silures, who lived in Wales during the Roman occupation of Britain.

Plate movements

A bird's eye view of the South Pole and changing distribution of the continents and oceans from the end of the Precambrian through the lower Paleozoic. Laurentia is North America plus Greenland, and Avalonia is southern Britain plus parts of western Europe.

Paleozoic plate movements

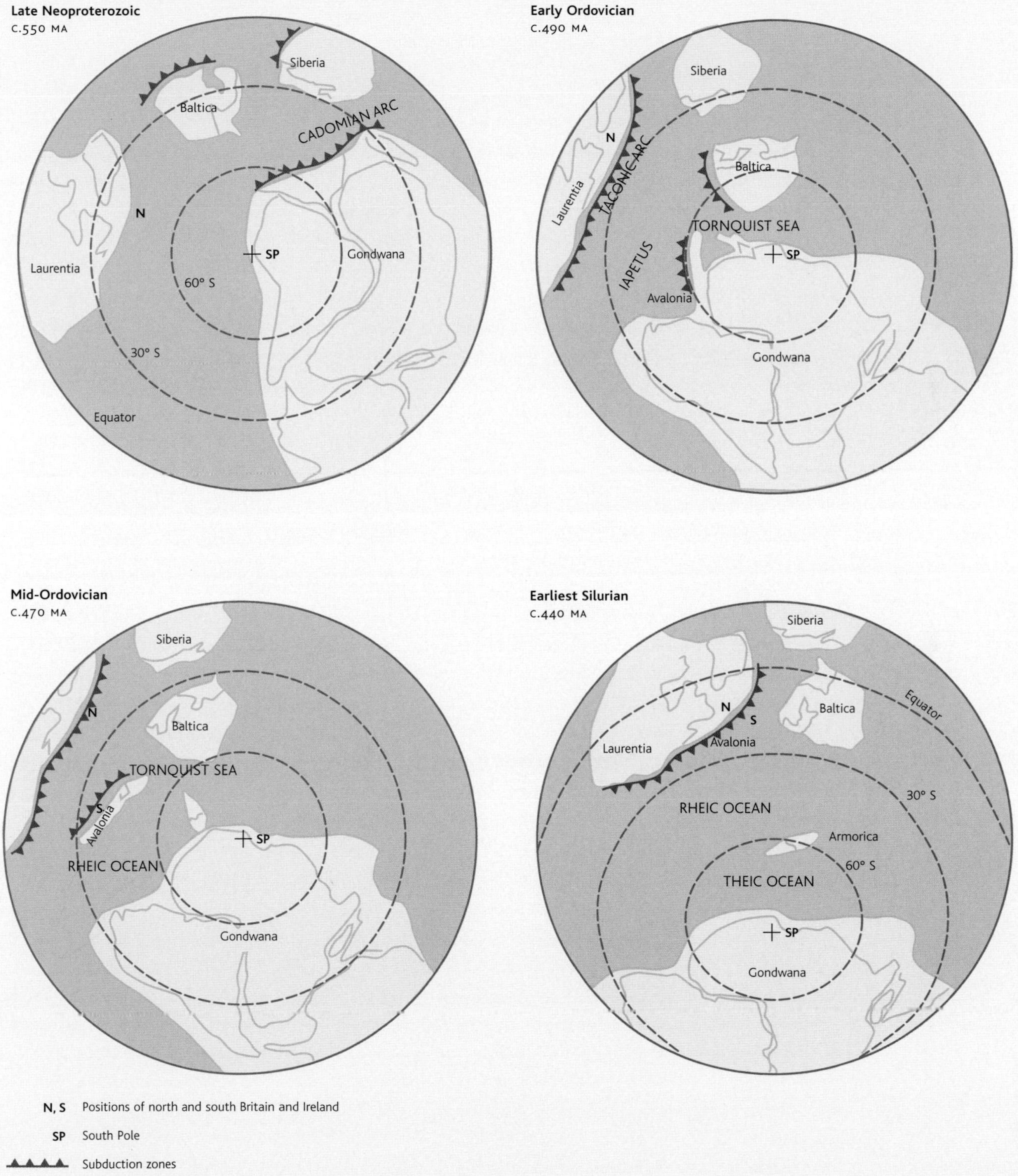

Late Neoproterozoic
C.550 MA

Siberia
Baltica
CADOMIAN ARC
N
Laurentia
60° S
SP
Gondwana
30° S
Equator

Early Ordovician
C.490 MA

Siberia
N
Laurentia
TACONIC ARC
Baltica
TORNQUIST SEA
IAPETUS
SP
Avalonia
Gondwana

Mid-Ordovician
C.470 MA

Siberia
N
Baltica
TORNQUIST SEA
S
Avalonia
SP
RHEIC OCEAN
Gondwana

Earliest Silurian
C.440 MA

Siberia
N   S
Equator
Laurentia
Avalonia
Baltica
RHEIC OCEAN
30° S
Armorica
60° S
THEIC OCEAN
SP
Gondwana

N, S   Positions of north and south Britain and Ireland

SP   South Pole

▲▲▲▲▲   Subduction zones

The move from sea to land

With the addition of Avalonia and Baltica, the enlarged continent of Laurentia was approaching the equator from the south in Silurian times, and it continued to move northwards with the rest of Gondwana closing up behind it. The fossil record of the Laurentian continent is luckily quite good because great thicknesses of sediments were shed by the weathering and erosion of the Caledonian mountain belt. Huge rivers carried great volumes of sand, silt, and mud across the landscapes into lakes and onwards to the sea which lay to the southeast of the British Isles and far to the north of Greenland.

During the early Devonian, these fresh waters were colonized by arthropods, molluscs, and numerous kinds of now extinct, strange-looking, armoured but jawless fish quite unlike anything alive today – although they are distantly related to the living lampreys. In fact, by late Devonian times, the arrival of jawed fish in the rivers and lakes was to change the course of evolution.

Overall, the landscapes were barren, devoid of soils and plants, except in low-lying swampy areas beside the rivers and lakes. The plants were all primitive and generally small (no more than 1m (3ft) or so high), and they were also leafless in early Devonian times. However, by late Devonian times, larger plants up to 20m (66ft) high with woody stems were forming the first swampy forests.

Generally, the Earth was in a greenhouse state, with high carbon dioxide and low oxygen levels – confirmed by stomatal densities in Devonian plants. All this plant evolution was confined, however, to low-lying areas where permanently wet conditions existed. Initially, the evolution of land plants (which began in late Ordovician or early Silurian times) depended upon fertilization in water. The evolution of spore-bearing plants such as the clubmosses (lycopsids), horsetails (sphenopsids), and ferns still required moist conditions for reproduction.

An extraordinary chance discovery in the 1930s by Danish geologists in late Devonian (363 million-year-old) sandstones from Greenland was to revolutionize our ideas about the evolution of backboned animals. After many years of painstaking preparation of the fossil bones from hard sediment, the Swedish palaeontologist Erik Jarvik (1907–1998) described the new animal that had been found as *Ichthyostega*. It was a 1m- (3ft-) long, salamander-like animal with four limbs, sideways flattened tail, and a broad, flat, bony skull with eyes on top. Jarvik saw the animal as a link between the fish and the first truly land-going amphibian tetrapods. He thought that it could both swim and climb out of the water onto the land. Thus the evolution of its limbs was seen to fit into the classic evolutionary picture of amphibians as an adaptation for life on land and transition between the fish and the reptiles. The story is not so simple, however, and is all the more interesting for its complications.

New fossils of both *Ichthyostega* and another somewhat similar tetrapod from Greenland called *Acanthostega*, plus other discoveries from the Baltic, Scotland, and eastern North America, all part of the Laurentian continent, have led to a radical revision of this scenario. Jenny Clack, a palaeontologist from the University of Cambridge, together with Michael Coates and Per Ahlberg, has shown that the lifestyle of the Greenland tetrapods was not an amphibious one, but rather that they were primarily water dwellers. They used their muscular, flattened tails for swimming in rapidly flowing rivers, while their forelimbs could perhaps grub around in sediment for food or be used for holding on to submerged vegetation; the hind limbs helped in steering the body. Of the two, *Ichthyostega* has the more robust limbs and long, bony ribs, and it may have been able to emerge from the water. However, new fossils now take the origin of limbs back even further, to 370 million years ago and the end of mid-Devonian times.

Per Ahlberg, a Scandinavian who now lives in Britain, had the good fortune to come across some early 19th-century fossils collected from Scotland. He also

Feet first
Around 450 million years ago a millipede-like arthropod with numerous pairs of small jointed legs and a long flexible and segmented body (about 1.5cm (½in) wide) crawled over a wet silt lake shore in what is today the English Lake District leaving a tell-tale trackway. The fossil remains of the animal have never been found but its trace fossil is one of the oldest known footprints of a land-living animal.

possessed the expertise to realize that what he was looking at was not just another lobe-finned fish, the group from which it has been speculated that the tetrapods first arose, but rather a new tetrapod. Using finds from the same locality that were scattered in various museum collections, Ahlberg pieced together *Elginerpeton*, as he called it, which shows a more primitive condition than seen in any of the other early tetrapods. It was about 1.5m (5ft) long, with paddle-like limbs clearly adapted for swimming and steering, but certainly not for walking. It also has features, especially in the long, pointed, and flattened skull, which link it to a group of lobe-finned fish called the panderichthyids. These 60cm (24in) long fish were originally discovered in Latvia, and as Ahlberg had worked on them he was able to recognize the connections. As the name "lobe-fin" suggests, these gilled fish had two pairs of stout,

muscular fins positioned like limbs, each with a single bone articulating to the shoulder girdle (the hind limbs and girdle are not known yet).

So the tetrapod limbs started off as pairs of muscular fins which were selected because they could assist a large, heavy fish in moving with sinuous body movements through shallow water. The development of more powerful limbs with hands, feet, and digits may have been a further adaptation which gave extra fitness to those animals that could dig for food and perhaps hold on in fast-flowing waters. Interestingly, it used to be thought by Jarvik that our five-finger/toe arrangement (technically known as a "pentadactyl limb plan") was originally secured in *Ichthyostega*. However, when Jenny Clack and her colleagues were preparing their specimens of *Acanthostega*, they found to their amazement that the animal had seven toes, and we

Greenland rocks

Some 360 million years ago, these rocks, which today are part of a mountainside (Stensio Bjerg) in eastern Greenland, were river sediments in the tropics inhabited by early fish, some of which (such as *Acanthostega*) had evolved four limbs as an adaptation for living in the fast-running waters and for grubbing about in the sediment for their food.

now know that it also had eight fingers. In addition, it now appears that *Ichthyostega* had eight fingers and we do not yet know how many toes.

Life on land is not easy compared to life in water. Bodies made of animal tissue are not very dense and, when immersed in water, especially seawater, they are buoyed up and supported. This is why the largest animals ever to have existed on earth, the blue whales, can do so well in the ocean. Movement in water is made relatively easy by using the friction of the whole body surface when thrown into muscular waves or by using fins or paddles.

Detection of sound and chemical "smells" is easy in water. Sight using eyes in clear, sunlit waters is also easy, and the eyes do not need lids, as their surface is

kept moist by the water. Also, obtaining oxygen is a relatively simple process of gaseous exchange across blood-suffused membranes (gills) in direct contact with oxygen-rich water. Finally, reproduction in water can be a very simple process, with gametes shed into the water where fertilization occurs, and larvae can take care of themselves right from the start.

By contrast, life on land is not much fun. The body is attacked by harmful ultraviolet light and desiccated by the dryness of the air. The eyes dry up, breathing is difficult and, without sound-detecting ears, hearing is impossible. Air is a light gas, and so the body is unsupported, making movement across land difficult without fairly complex biomechanics. On top of all that, reproduction in the manner of fish is impossible. So it is little wonder that we use the expression "like a fish out of water" to describe a helpless and hopeless situation.

The transition from life in the water to successful life on land required so many biological changes, which had to be in place in advance, that it is somewhat surprising that it happened at all. There must have been a considerable advantage for some life forms to make the move from water to land. It is possible to understand how some of the preadaptations were selected by the lobe-fins while others were selected by the early aquatic tetrapods. Much of the selection pressure was coming from the rapidly changing environments in which the animals were living. Rivers and lakes with aquatic plants and well-vegetated banks provide a great variety of ecological niches in which different kinds of water-dwelling creatures can survive. The problem is that, on the geological time scale, these environments are essentially ephemeral, constantly changing and eventually disappearing altogether.

Rivers shift their channels, meanders form, and oxbow lakes are cut off from the mainstream. Water bodies can vary from being swift flowing and well oxygenated to being stagnant, anoxic pools filled with rotting vegetation which subsequently dry up. The competition for food and oxygen can become intense. Reproduction is particularly hazardous in that the fertilization of large numbers of eggs and, from them, defenceless developing larvae can simply be providing another creature with a free meal. Being able to gulp oxygen directly from the air was a considerable advantage and a trick evolved in earlier Devonian times by the first lungfish. Stout, muscular fins which could aid rapid movement through shallow, muddy water was another advantage. A large, multi-toothed, gaping mouth and predatory habit were also advantages.

An interesting problem is that clues about how animals reproduce are not well preserved in most fossil remains. As a result, we do not know exactly how early land-going tetrapods managed the reproductive process, but it is almost certain that they were similar to living amphibians and had to return to the water to breed. Only with the evolution of the amniote egg in the first reptiles did they become fully independent of the water. The group to which living amphibians – frogs, toads, salmanders, etc (the lissamphibians) – belong is a more recent "invention", but was preceded by a number of extinct amphibians which evolved in Carboniferous times and some of which were very large, crocodile-like predators. The term "tetrapod" is used for these early four-limbed creatures because they have a mosaic of fish-like and more evolved characteristics and thus cannot be assigned to the strict biological grouping we know as amphibians. The last decade or so has seen the discovery of a whole succession of new fossils, and old ones languishing in museum collections have been reinterpreted. Together they are providing new data which promises to make the early evolution of the tetrapods much clearer.

Pederpes
This small (15cm (6in) or so) four-limbed vertebrate called *Pederpes* is the oldest known animal with limbs fully adapted for walking on land, and was found in early Carboniferous (Mississippian) strata in Scotland.

Putting the backbone into life

Those tumultous early decades of geological discovery at the beginning of the 19th century revealed that, when all the sections of fossiliferous rock strata exposed at the surface were joined together sequentially, they amounted to many tens of kilometres in thickness. As we have seen, there was a critical debate over how life was distributed throughout these strata. At the beginning of the century, Charles Lyell was arguing that fossil representatives of most major groups of organisms could be found within even the oldest fossiliferous strata. Even in the 1820s and 1830s, it was evident that fish fossils could be found in ancient Old Red Sandstone strata (now known to be of Devonian age), and mammals were present in Jurassic strata. If this were true, then perhaps it would be possible to detect in the fossil record the moment of creation of life at some remote period in the past. By the mid-19th century, however, when magnificently fossiliferous ancient strata in Scandinavia, Bohemia, and New York State were becoming better known, it seemed to be that the most ancient fossiliferous strata did not contain more advanced organisms.

In his multi-volumed and beautifully illustrated description of the fossils of Bohemia, Joachim Barrande (1799–1883), showed that the earliest stages of Murchison's Silurian system contained trilobites, brachiopods, and extinct groups of cephalopods. The trilobites are an extinct group of arthropods; the brachiopods are a group of marine bivalved shellfish which look somewhat like clams, but are biogically separate; and the cephalopods were squid-like animals which had protective conical shells, often coiled. Gradually it became evident that these fossils actually belonged in Sedgwick's Cambrian system, but there was no evidence of backboned animals among them.

Thanks to recent discoveries in early Cambrian rocks in China and some earlier finds in the somewhat younger Cambrian deposits of Canada, we can now trace the origin of backboned animals into early Cambrian times. Since the end of the 19th century, biologists have known a great deal about primitive vertebrate animals thanks to the existence of a number of surviving groups which preserve various aspects of this evolutionary phase. The fish might be thought to be pretty simple vertebrates with their simple backbones, laterally flattened bodies and muscular sideways flexing for swimming, plus gills for breathing etc. But what we tend to think of as typical fish, for example, the salmon, are in fact advanced bony fish (osteichthyan teleosts, to use the technical terms) which evolved within the past 100 million years or so (Cretaceous times). Primitive teleosts, however, originate back in Triassic times, and the large group of osteichthyans are now known to have evolved in early Silurian and perhaps late Ordovician times.

Biologists have for a long time considered that the cartilaginous fish, mainly familiar to us today as sharks and rays, and which are in many ways very different from the bony fish, are more primitive. There are good biological reasons to think that they probably evolved before the bony fish, but the problem with proving this is that cartilage does not usually fossilize, as it generally has no mineral content. But we also have some survivors of even more primitive fish-like animals, the lampreys and hagfish. These are not particularly familiar animals to most people, but they are of great biological and evolutionary interest. Both are jawless. The lampreys are specialized freshwater ectoparasites (parasites that live on the outside of the host organism) to other fish, while the hagfish, or slimehags, are sea-dwelling scavengers whose acute sense of "smell" allows them to detect corpses from a great distance. Underwater filming has shown that hagfish are extremely efficient at rapidly removing all the tissue from cadavers, even those as big as a whale lying on the ocean floor.

Biologically, the lampreys and hagfish are quite different in detail from one another, but there is a greater difference between all the other fish and these two groups combined. According to the molecular clock, they originated some 750 million years ago; however,

again, because their internal skeleton is cartilaginous, their fossil record is not very good. The hagfish record only dates back to the Carboniferous, while the lampreys extend back to the late Ordovician. Their most important distinguishing features include a structural support for the gills which has the gills on the inside, and the absence of jaws whereas in both the cartilaginous chondrichthyans (sharks and rays etc) and the bony fish, the gills are on the outside of the supporting structure.

The classical idea of the evolution of the jaws sees them as evolved from the skeletal elements of the first gill support arch. As the mouth enlarges and extends backwards, the next gill arch becomes modified as a supporting strut to prevent the jaws from being pushed back and to help brace them against the brain case. The problem with the living lampreys and hagfish is that they are thought to have modified considerably from their original form as they have taken up parasitic and scavenging modes of life.

The fossil record does, however, preserve a large number of extinct jawless "fish" groups which thrived in the sea from late Cambrian to late Silurian times, when they invaded fresh waters, and continued to flourish to the end of the Devonian, when they all became extinct. Many of them are quite bizarre, with the head and thorax covered with a thick, tough, leathery "armour" of mineralized, bone-like plates covered with a thin layer of skin. In some, there was a pair of movable pectoral fins, and, in most, only the tail region was flexible, although still covered in scales. Their diversity of form shows that they had adapted to occupy a wide range of ecological niches, but the lack of jaws and teeth shows that they must have all been either filter feeders or microphagous feeders – ie consuming soft food such as organic-rich muds (detritivores) or soft plant material. Abundant mud-filled faeces that are associated with the occurrence of some groups confirm that at least some of them were indeed detritivores. Some of these fossil agnathans had the head region covered with a single bony carapace-like head shield that fossilizes well. Dissection and moulding of the internal spaces in these head shields have provided a great deal of information

Fishy ancestors

In mid-Devonian times, around 370 million years ago, freshwater lakes and rivers were teeming with diversifying groups of early fish. Among the more advanced groups were the rhipidistians, such as *Osteolepis macrolepidotus* (from Orkney in Scotland, seen here), with a broad head and tapering body with two pairs of fins and covering of large, bony scales.

Ancestor of the vertebrates

Among living vertebrate-related animals exists a fascinating little creature, the lancelet, technically known as *Branchiostoma* (it used to be called (*Amphioxus*). It is some 4cm (1¾in) long, laterally flattened like a slender leaf, and pointed at both ends, so it can be difficult to tell the head from the tail at first. Indeed, the lancelet can swim both forwards and backwards, and lives partly buried, tail first, in sea-bed sands. It feeds by filtering tiny organic particles from the water. Water is sucked in through its mouth, passes over and through a barred sieve-like structure in the throat region, and travels out of the body. This way, the lancelet obtains both oxygen and food, which are removed from the water by tissues covering the sieve structure. It is possible that some of the fossil agnathans were similar filter feeders. Otherwise, there is not much to the head of the lancelet except some photoreceptor cells which act as crude "eyes" and sensory tentacles surrounding the "mouth".

What is particularly interesting about the lancelet is its basic anatomy. There is a flexible stiffening rod called a notochord which runs from the very tip of the head to the end of the tail, encased in two side-by-side and parallel segmented series of muscle blocks. When these muscle blocks contract in waves passing from the tail forwards, the body bends from side to side and moves forwards through the water, the same basic mechanism as seen in most fish. Reversing the contractions sends the animal backwards. Above the notochord lies the dorsal nerve cord. Study of the embryological development of agnathans and true fish shows that the notochord is the precursor to the vertebrate backbone. So this elongate axial stiffening rod is a core developmental preadaptation for the vertebrate body form and mode of life.

The critical feature that differentiates the chordates from the vertebrates is seen in the development of the neural crest cells within the embryo. In the vertebrates, some of these cells break away to form structures such as the eyes, skull support structures (skull), and head muscles. Although the lancelet does not have a true neural crest, it does have cells in a similar position which express similar genes.

Burgess Shale (far right)
High on a mountainside in British Columbia, Canada, the World Heritage Site of the Burgess Shale preserves one of the most diverse and remarkable of early fossil faunas of mid-Cambrian age, some 505 million years old. Like somewhat older Chinese strata, these marine deposits preserve not only a wide range of invertebrate shellfish but also some of the earliest chordates.

Before the fish (below)
The living lancelet *Branchiostoma* is a primitive but still successful fish-like marine animal that grows to 7.5cm (3in) long and has a laterally compressed, tapered body, within which are the basic structures of vertebrate organization such as a flexible rod (the notochord), a dorsal nerve cord, and a sieve-like structure in the throat region. Lancelets are chordates and are remarkably similar to some of the earliest fossil chordates.

about the nerves, blood vessels, brain structure, and gill form which confirms their affinities with the lampreys, which still exist today, in particular.

Some of these extinct agnathan groups were not armoured and looked much more like small "fish" (up to 20cm (8in) long) with laterally flattened bodies covered in small scales, but scales which are structurally and compositionally different from the scales of bony fish. These agnathans swam like most modern fish do – by throwing their bodies into lateral waves using muscle blocks which are serially repeated along the length of the body. The tail has both upper and lower lobes, but these are often of different sizes, which would have pitched the animals either head down or head up, suggesting that they were either water-surface or sediment-surface feeders. Lateral fins, often present as continuous lateral fin folds, helped to stabilize forward motion for normal swimming. The gill openings were numerous (eight to 13), the mouth was circular and right at the front of the head, and the eyes were often quite large. Clearly, these were active and possibly quite fast-swimming animals. Some of them had skins without scales, but our record of these forms is extremely poor. Consequently, not much is known about the important transition from jawless to jawed fish that occurred sometime from the end of the Cambrian and the early Ordovician period. But we can go another step backwards in time.

When we look around at some other primitive animals still alive today, it turns out that there are a few groups (such as the sea squirts and pterobranchs) which have a notochord, but only in the larval stage. More interesting from the evolutionary point of view are the extinct conodonts. These were very successful and abundant marine creatures which evolved in late Cambrian times and became extinct at the end of Triassic times. They were generally small (4cm (1¾in) or so long), but one giant group (40–80cm (16–32in) long) did evolve in the cold waters of late Ordovician times. Only recently has any fossil information about their body form been found – most of their fossil remains consist of tiny barbed bars, barely visible to the naked eye. Made of calcium phosphate, they look like tiny but sharply pointed teeth, and indeed they were first described as such by the Russian palaeontologist

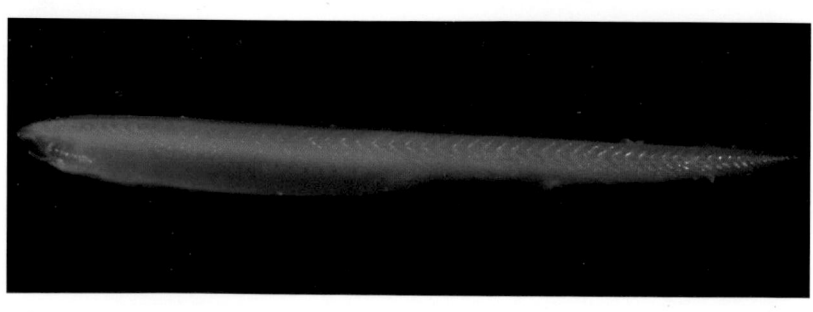

Christian Pander (1794–1865) in 1856. After nearly 150 years of confusion, the discovery of a group of conodont teeth found at the end of an elongate, worm-like body allowed them to be reassessed taxonomically and biologically.

Several pairs (mirror images) of mineralized conodont elements together make up a toothed feeding apparatus at the front of the animal, which also has paired vertebrate-type eyes and a notochord with segmented blocks of muscles on either side. Debate is still ongoing about the exact status of the conodonts, but they are the earliest vertebrate-related animals to have mineralized teeth. They were active predators, but consumed only tiny organisms and could not attack larger creatures. Indeed, they were probably wiped out by the evolution of some of the predatory bony fish. Their presence in late Cambrian rocks shows that there was a diversity of early chordates (animals with a notochord, but no true segmented vertebral column) the interrelationships of which are still not entirely clear.

Most recently of all, some curious little lancelet-like fossils have turned up in China which represent a considerable advance on the lancelet condition. *Yunnanozoon* was first described in 1995, and it was soon followed by three others, one of which, *Haikouella*, is represented by more than 300 specimens. Unlike *Yunnanozoon*, *Haikouella* seems to have a simple ventral heart, dorsal and ventral aortic blood vessels, gill filaments, a distinct head with forward swelling of the dorsal nerve cord (a primitive brain?), and possibly eyes. The newer finds preserve skull-like structures, gill arches, fin supports, better defined eye capsules, and a large heart behind the gills, a combination that places *Myllokunmingia* (meaning "the fish from Kunming") and *Haikouichthys* (meaning "the fish from Haikou") as primitive vertebrates.

All this new evidence suggests that our remotest backboned ancestors were actively swimming animals, with well-developed sense organs (eyes and perhaps chemoreceptors) placed at the front of an elongate

Haikouicthys

An astonishing variety of soft-bodied fossil organisms has been exceptionally well preserved in early Cambrian seabed deposits of Chengjiang in Yunnan Province, southern China (some 530 million years old). They include *Haikouichthys*, claimed to be one of the earliest known vertebrate fossils, preserving not only the notochord but also true gill arches and perhaps cranial cartilages in the head.

The evolution of early vertebrates

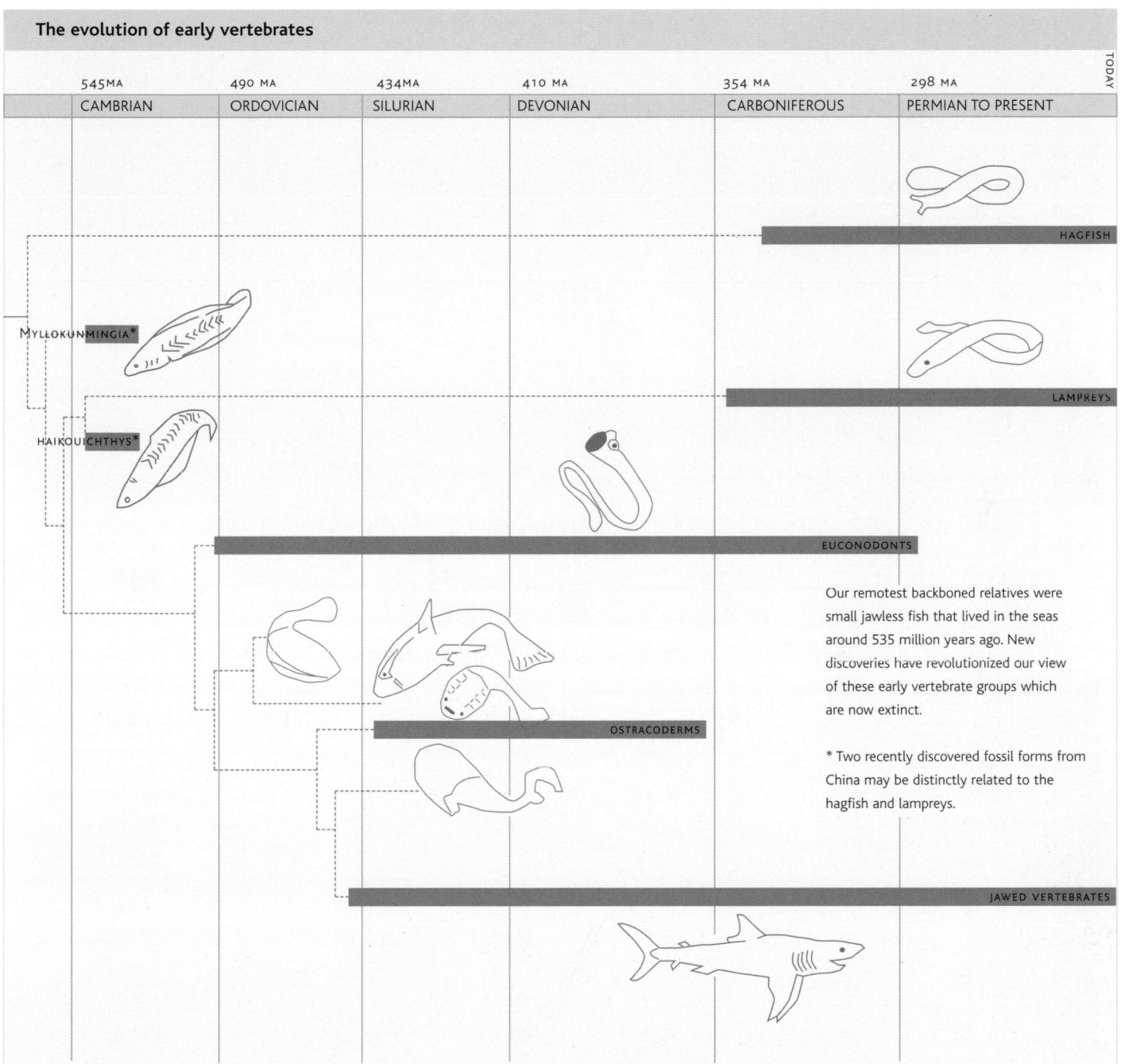

545MA	490 MA	434MA	410 MA	354 MA	298 MA
CAMBRIAN	ORDOVICIAN	SILURIAN	DEVONIAN	CARBONIFEROUS	PERMIAN TO PRESENT

TODAY

HAGFISH

MYLLOKUNMINGIA*

HAIKOUICHTHYS*

LAMPREYS

EUCONODONTS

Our remotest backboned relatives were small jawless fish that lived in the seas around 535 million years ago. New discoveries have revolutionized our view of these early vertebrate groups which are now extinct.

OSTRACODERMS

* Two recently discovered fossil forms from China may be distinctly related to the hagfish and lampreys.

JAWED VERTEBRATES

body, coordinated by a developing anterior brain, which was protected by a non-mineralized "brain-box", or "skull". The later Cambrian conodont animals perhaps represent the first vertebrates with mineralized tissues. As we have seen, this was not in the form of skeletal tissue, but rather teeth allied to a predatory habit.

Furthermore, it is claimed that *Myllokunmingia* is closer to the hagfish and *Haikouichthys* to the lampreys.

If true, this would indicate that these two vertebrate groups had already split by early Cambrian times and must have shared a common ancestor within the late Precambrian. Such an early ancestry is supported by molecular clock evidence derived from measures of the genetic distance between living lancelets, the hagfish, and the lampreys, which dates a common vertebrate ancestry at around 750 million years ago.

7
Life's beginnings

Our view of the prehistory of life is still very biased towards what has happened over the last 545 million years since the beginning of Cambrian times and the appearance of abundant fossils in the rock record. We now suspect that life probably began soon after the first primitive atmosphere and oceans had formed on Earth, around 4,200 million years ago. However, the whole business had to start again because a catastrophic late bombardment of the Earth by meteorites around 3,900 million years ago is thought to have led to a complete meltdown and destruction of any initial atmosphere, oceans or life. Nevertheless, recent discoveries and analyses of what are known as chemical fossils – hydrocarbon nodules – suggest that by 3,800 million years ago life had probably become re-established. The early atmosphere and oceans were very different in their chemistry from those of today and only those single-celled micro-organisms that could tolerate the conditions, especially the lack of oxygen, would have evolved.

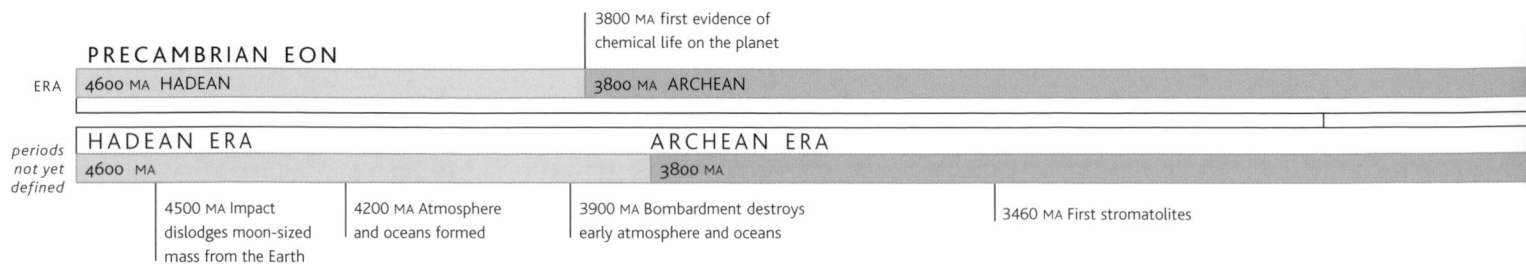

	3800 MA first evidence of chemical life on the planet	
PRECAMBRIAN EON		
ERA	4600 MA HADEAN	3800 MA ARCHEAN

periods not yet defined	HADEAN ERA	ARCHEAN ERA	
	4600 MA	3800 MA	
	4500 MA Impact dislodges moon-sized mass from the Earth	4200 MA Atmosphere and oceans formed	3900 MA Bombardment destroys early atmosphere and oceans	3460 MA First stromatolites

Scientists are actively investigating similar extreme environments today where temperatures are either very high or low and where the water is either very acid or very alkaline, to see what kinds of microbial life can tolerate such conditions and how they do it. The aim is to try and gain some insight into the kind of life forms which might have first evolved.

In recent decades scientists have also tried to replicate the chemical and physical conditions of the early Earth to see if they could generate life from natural combinations of organic chemicals and physical circumstance which might have "sparked" the inherent energy that characterizes living from inanimate matter. Although in some ways successful, these were still very naive and crude experiments but they did force a re-examination of what was known about the basic biology of life, which was very little. The result was a spectacular boom in fundamental research into the workings of the most primitive cells, their chemistry, and molecular biology.

Ancient sediments

Although metamorphosed by heat and pressure, folds and faults, sedimentary lamination is still visible in these late Precambrian rocks. It was from such evidence that 19th-century geologists realized that sedimentation processes seemed to extend back in time indefinitely, beyond the base of the Cambrian.

1200 MA The first multi-celled organisms date from the middle of the Proterozoic period

610 MA The first large marine animals appear

PHANEROZOIC EON

2500 MA PROTEROZOIC | 545 MA PALEOZOIC | 248 MA MESOZOIC | 65 MA CENOZOIC

PROTEROZOIC ERA

2500 MA

700 MA First oxygen produced ̷ photosynthesizing microbes

2300 MA First snowball Earth

1200 MA First multicellular algae

720 MA Snowball Earth

580 MA First Ediacarans; Doushanto fossil embryos

555 MA First shelled organisms

The discovery of the Precambrian

Suilven, Scotland
Glacially eroded strata of late Precambrian horizontally bedded sediments (Proterozoic, around 980 million years old) lie on more ancient, highly metamorphosed Precambrian rocks (Archean, dating back to 2,800 million years ago) of the surrounding landscape.

By 1835, when Adam Sedgwick first defined the Cambrian system of rock strata in Wales, the foundations for the division of geological time, based on sedimentary strata and their contained fossils, were literally laid down in tablets of stone. Sedgwick and his geological colleagues were well aware that the testimony of the rocks did not begin with the base of the Cambrian, although they did believe that the story of life did. In the early decades of the 19th century, the story of life seemed to stretch back to a beginning with the first fossils found in Cambrian strata. Beyond that, there were still plenty of "deeper" sedimentary rock strata which were clearly pre-Cambrian; however, they seemed to be devoid of life and were consequently called Azoic (meaning "without life"). In many places, even older but highly deformed and metamorphosed

strata could be seen, often intruded by igneous rocks – the Primitive rocks of Werner's and other early classifications.

By the mid-19th century, great tracts of apparently Precambrian metamorphic rocks were being found in Scotland and on a much greater scale in North America, where the Canadian Shield was turning out to include a truly vast area of such rocks. Some of these looked as if they had been bedded sedimentary strata before they were metamorphosed. There were, however, problems with trying to unravel Precambrian geological history. As there was no independent method of dating the rocks and no fossils, all that could be done was to try to establish their relative stratigraphic position, which was generally very difficult in highly deformed rocks. Yet, as was being shown in the Scottish Highlands, it was possible to do this. The problem was that intense metamorphism destroys most fossils, so that it could not be assumed that all metamorphic rocks were Precambrian – and, indeed, they are not. In fact, it was to turn out that quite a lot of the Scottish metamorphic rocks are lower Paleozoic in age.

We now know that nearly all the major continents, eg. Laurentia (North America and Greenland), and Amazonia (northern South America), have significant areas of ancient Precambrian exposed at the surface, with younger rocks and often mountain belts wrapped around them. Many of these Precambrian rocks are highly deformed, but not all, and many contain economic mineral resources of global importance, such as the banded iron formations of Hamersley, Western Australia; Isua, western Greenland; Namibia and Transvaal, southern Africa; Siberia; Brazil; and Lake Superior, Labrador, and Minnesota in North America. Their ages range from 3,850 to 700 million years old, with the bulk being between 2,750 and 1,800 million years old. The Hamersley reserves are estimated to be around 27 million tonnes with more than 55 per cent iron in the ore.

Darwin suspected that there should be fossils of primitive life forms to be found in Precambrian rocks. The Scots-Canadian geologist J W Dawson (1829–1899), who had shown Lyell the Joggins fossil trees, made the first breakthrough in 1865. Dawson found some organic-looking remains in Precambrian limestones exposed

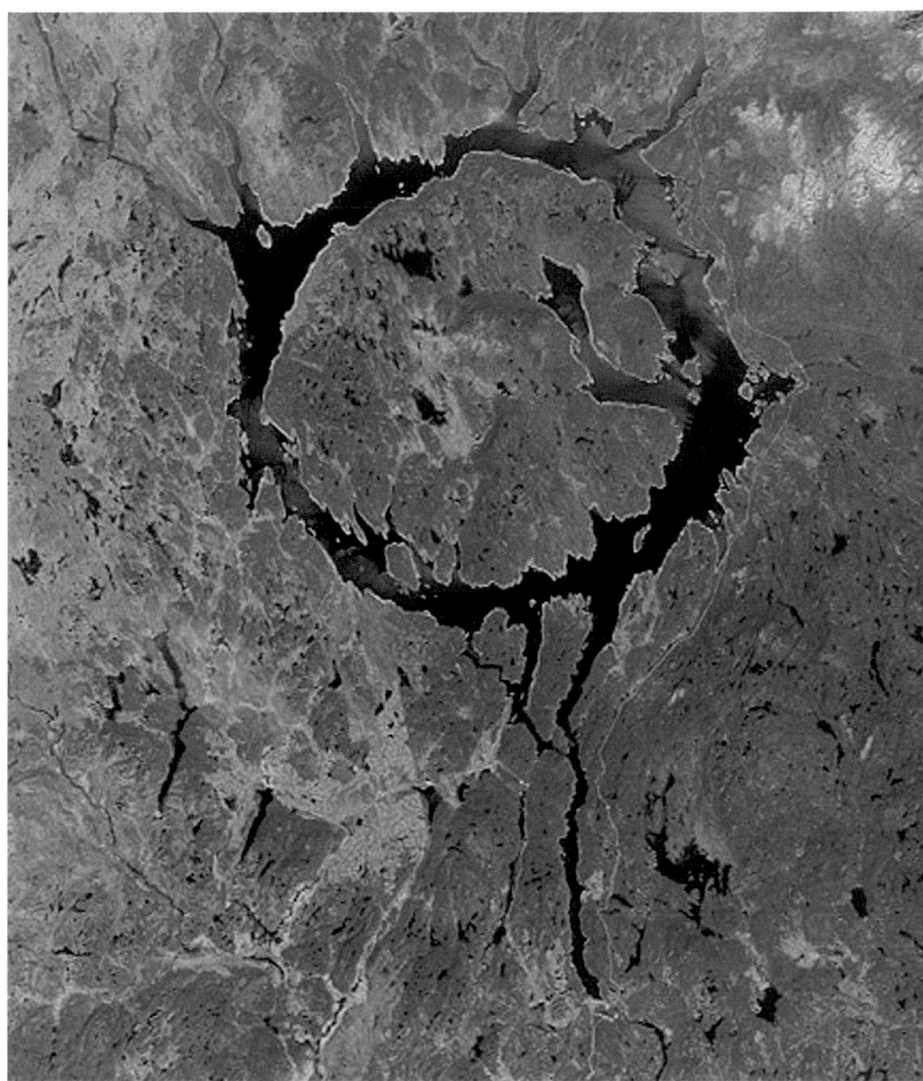

along the Ottawa River near Montreal, Canada and described them as the fossil of a giant, unicellular organism which he called *Eozoon canadense*, meaning "dawn animal from Canada". Dawson's reputation was such that his fossil was generally accepted as the first genuine evidence for life extending back into the Precambrian.

Although it subsequently turned out that Dawson's *Eozoon* was not a fossil at all, but an inorganic mineral growth, the conceptual breakthrough had been made. Geologists now expected to find Precambrian fossils if they looked hard enough in the least deformed of these most ancient sedimentary rocks, and it was not long before they did. By the end of the 1880s, a rising star of American palaeontology, Charles Walcott (1850–1927),

Canadian Shield

At 70km (43½ miles) in diameter, the 220-million-year-old Manicougan impact crater, is still clearly visible indenting the much more ancient Precambrian strata of the Canadian shield. Deep glacial erosion by Quaternary ice sheets has scoured the rocky landscape and infilled the remaining circular depression with melt water.

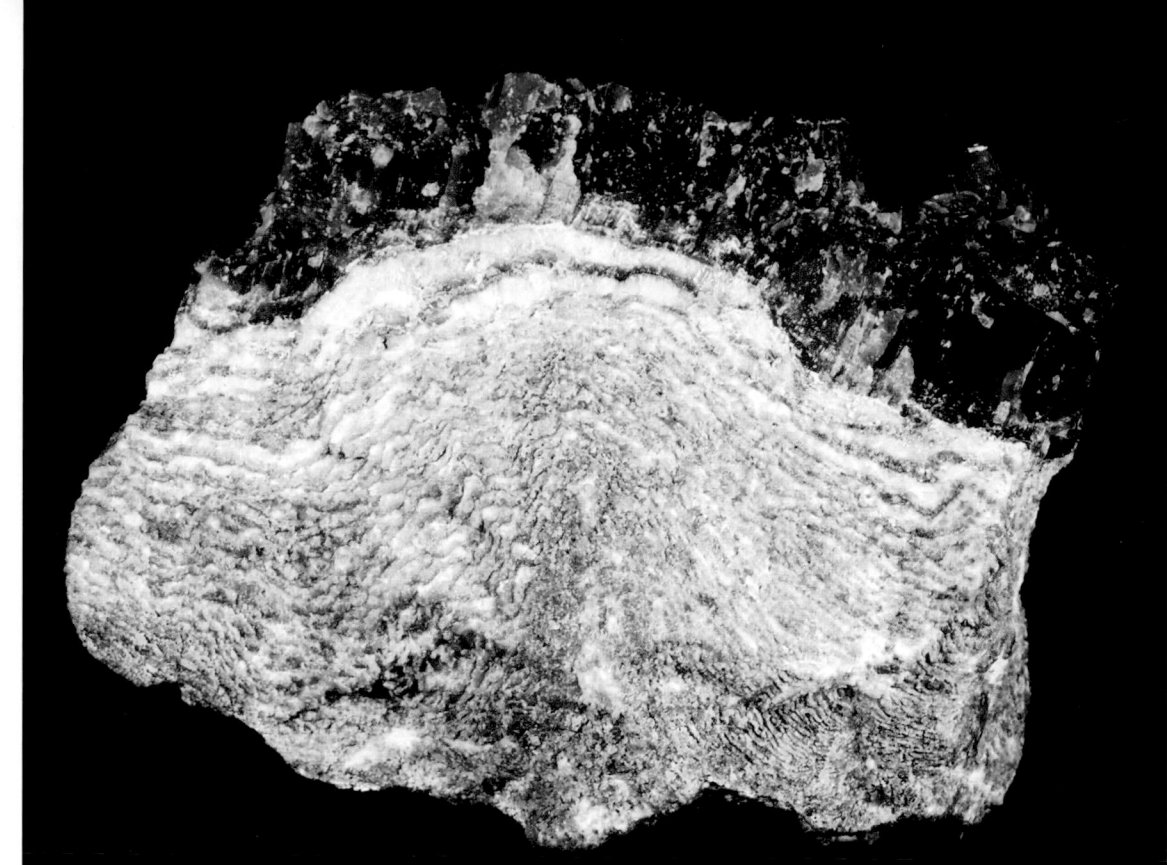

Eozoon

This cut and polished section through a Precambrian rock from the Canadian Shield was thought to represent the oldest known traces of fossil life and was first described by JW Dawson in 1865. Although its bulging laminae do look organic, more mineralogical detailed analysis proved otherwise.

had found some strange, organic-looking structures in Precambrian strata at the bottom of the Grand Canyon. By 1891, Walcott was convinced that since "life in [Precambrian] seas was large and varied and there can be little, if any doubt that it is only a question of search and favourable conditions to discover it".

By the end of the 19th century, enough was known about the processes of fossilization for it to be realized that it is very selective and generally only preserves organisms with hard parts. As primitive life would lack the necessary hard parts to fossilize and also be very small, if not actually microscopic, the chances of any such organisms being preserved within rocks seemed very slim. Soft-bodied organisms such as jellyfish are very rarely fossilized anywhere in the fossil record, and then only in special circumstances. Microscopic soft-bodied organisms stand virtually no chance of being preserved unless by some process of secondary mineralization.

It was not until the 1950s that such fossils were found – the debunking of Dawson's *Eozoon* had set back the unrewarding task of looking for Precambrian life. Again it was in Canada that the real breakthrough was made, when Stanley Tyler (1906–63), an American economic geologist, found some stunningly well-preserved microfossils in ancient Precambrian Gunflint

ironstones in Ontario which are exposed along the shores of Lake Superior. The ironstones occur as bedded sedimentary ores with layers of chert, some of which are formed in distinctive mound-like structures, similar to those from the Grand Canyon described by Walcott as *Cryptozoon*. At the time, they had no idea of the age in years of the rock, but they did know that it lay deep within the Canadian Precambrian succession. It has now been confirmed as truly ancient, with a radiometric date of 2,100 million years.

It turns out in retrospect that, around the same time, a Russian scientist, Boris Timofeev (1916–82), was discovering and describing similar microorganisms from Precambrian strata in the Urals. As his discoveries were published in Russian journals and this was the period of the Cold War, his pioneering work did not become known about in the West for another decade or so. With the Gunflint discovery, however, it was realized that, when certain special conditions of preservation prevail, microscopic cells that are unprotected by shells or skeletons will fossilize, and so the race began to find just how far back into Precambrian times life extended. The additional breakthrough that helped the whole enterprise enormously was the advent of radiometric dating in the 1950s, again aided by wartime technology associated with work on nuclear fission undertaken by the Manhattan Project. The discovery that Precambrian time amounted to 4,000 million years, some eight times longer than the Phanerozoic with its relatively abundant fossil record, was something of a shock, but it also provided an enormous *terra incognita* for research. Being able to develop some chronology and date rocks more exactly were pivotal to this.

We now know that there are chemical traces of life found in rocks from Greenland as old as 3,850 million years. There is probably a much better chance of preserving such chemical signs of early life than there is of body fossils. However, the nature of such early fossils is hotly debated, and the desire to be the discoverer of the oldest fossils on Earth is as strong as ever it was since the stratigraphic empire-building days of Sir Roderick Murchison.

At the other end of the Precambrian, large organisms had evolved by 600 million years ago, although nobody is entirely sure about what kind of organisms they actually were. They are known as the Ediacarans, after the Australian locality where well-preserved fossils of them were first found. In addition, we now know that the race towards building protective shells also began around 560 million years ago, just before the beginning of Cambrian times.

The whole span of Precambrian time is over eight times longer than the Phanerozoic, which is that part of prehistory we are most familiar with, the last 545 million years. Technically, it has been superceded by major divisions from the most ancient, the Hadean (4,600–3,800 million years ago), the Archean (3,800–2,500 years ago) and the Proterozoic (2,500–545 million years ago). Already, the Proterozoic is subdivided into the Paleo-, Meso- and Neoproterozoic, and as more is learned about the details of Precambrian history, no doubt further subdivisions will be introduced.

Ediacaran homeland
The Pound Quartzite, a late-Precambrian marine sandstone, in the Ediacaran Hills of the Flinders Range, Southern Australia was found by Australian geologist Reginald Sprigg to contain well preserved fossils. They represent a variety of soft-bodied organisms which were thought to be related to jellyfish.

The search for the oldest fossils

The oldest fossils visible to the naked eye are mound-shaped structures known as stromatolites and are around 2,720 million years old from the Tumbiana strata of Australia. Stromatolites were first found and described by an eminent American earth scientist, James Hall (1811–98), who was chief palaeontologist of New York State Geological Survey and then director of the Natural History Museum in Albany. In the mid-19th century, he described and beautifully illustrated in remarkable detail the Paleozoic fossils of New York State, and it was Hall who first sent Walcott out to investigate curious mound-shaped structures in Cambrian limestones at Saratoga, New York State. Hall named these structures *Cryptozoon*, meaning "hidden life", and we now know that they are stromatolites.

Stromatolites are essentially layered mounds of sediment generated by the interaction of bacterial films with the deposition of sediment on shallow, warm-water sea beds. We now know a lot about them

Gunflint Chert

In the 1950s, American geologist Stanley Tyler found that some late Precambrian sedimentary strata exposed along the coast of Lake Superior, known as the Gunflint Chert, contained remarkably well-preserved micro-fossils, such as this organic walled sphere, called *Eosphaera tyleri* (early Proterozoic in age, around 1,878 million years old). They were among the first genuine Precambrian fossils to be described. Their discovery led to a renaissance in the search for ancient life.

because, surprisingly, they are still with us, thriving in the tropical seas off Western Australia and the Gulf of Mexico. Biologically, they are very interesting and very important for our understanding of how life evolved. Essentially, stromatolites are single-celled microorganisms, the genetic material of which is not contained within a discrete membrane-bounded nucleus, but scattered throughout the cell. This primitive condition is known as prokaryotic, and some prokaryotes such as the bacterial organisms which build stromatolites produce oxygen through photosynthesis. To do this, they need access to strong sunlight and therefore live in shallow water.

Bacterial films grow as separate mats over the sea-bed sediment, but currents and wave activity tend to wash sediment over them. The bacterial filaments grow up through the sediment to gain access to the sunlight again and, as the process is repeated, each mat grows into a laminated mound. The mounds vary in size from being a few centimetres wide up to a metre or so wide. Their height depends on the local environment of sedimentation. In Shark Bay, Western Australia, the mounds are 10–50cm (4–20in) wide and grow up from the seabed to heights of 1.5m (4¾ft) or so.

The bacteria die soon after being covered up, and generally all trace of their cellular tissue is soon lost. The distinctive laminated, mound-shaped masses of sediment, however, are more difficult to destroy, especially if they are buried where there is active deposition.

Precambrian fossil stromatolites several metres high have been found. Their production of oxygen was vital in gradually changing the Earth's atmosphere from being dominated by carbon dioxide, nitrogen, and small amounts of hydrogen to one that was more hospitable to the evolution of a greater diversity of life. By 2,000 million years ago, oxygen levels had risen to some 15 per cent of their present level, and it was not until some 600 million years ago that it may have approached present levels.

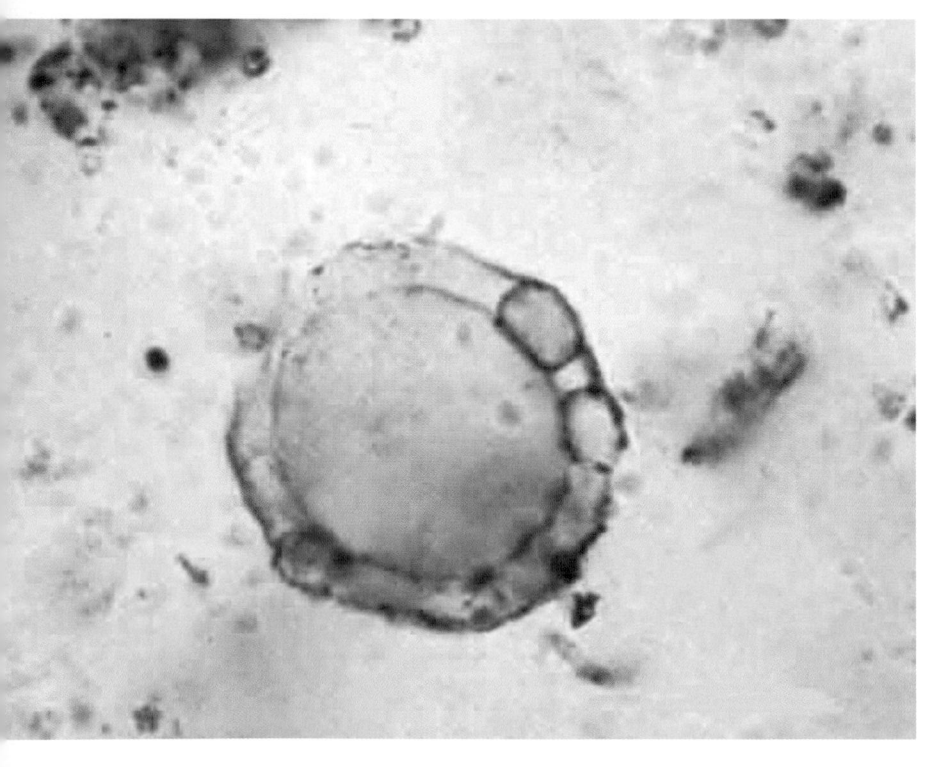

The very oldest known fossils are preserved in a different way to the stromatolites, however, and they are known as "chemical" fossils. All that remains are the traces of organic carbon compounds, complex molecules preserved as tiny hydrocarbon residues in what were originally sedimentary rocks, but are now metamorphosed. These fossils have been found at Isua in western Greenland and are dated at about 3,850 million years old. Chemical analysis of the molecules shows that they are so complex that they must have been derived from bacteria-like organisms, suggesting that the origin of life itself must have predated them, perhaps as long ago as 4,000 million years. Interestingly, the discovery of oil and gas in 2,000 million-year-old Precambrian sediments from Siberia and Southern Africa shows that life in the oceans must have become abundant by that time, although it was all microscopic.

The big evolutionary innovation which came around 2,100 million years ago was the development of more complex cells, called eukaryotes, through enclosure of the cell's genetic material within a membrane to form a nucleus. The first of these were still microscopic unicells which lived in surface waters of the seas and oceans, and obtained their energy through photosynthesis. We now have direct fossil evidence that, by 1,200 million years ago, multicellular organisms had evolved, although it is likely that the origins of multicellular organisms are much older than this. The fossils are small, seaweed-like red algae, the cells of which clustered together to form filaments.

Recently, Nick Butterfield, a University of Cambridge palaeobiologist, has found and described one of these red algae from Precambrian rocks in Arctic Canada, which belongs to a surviving group called the "bangiomorphs". Its cells show patterns of cell division which are the typical product of sexual reproduction. This is the oldest evidence for sex in which there was a mixing of genetic products from different individuals. The advantage of sexual reproduction is that it promotes a much greater genetic mix in offspring and hence variation and diversity of individuals. Without sex, it is likely that life on Earth would have consisted of not much more than oceans full of microscopic unicells, slimy multicellular filaments, and sheets of asexually budded monotonous clones of the parent cells.

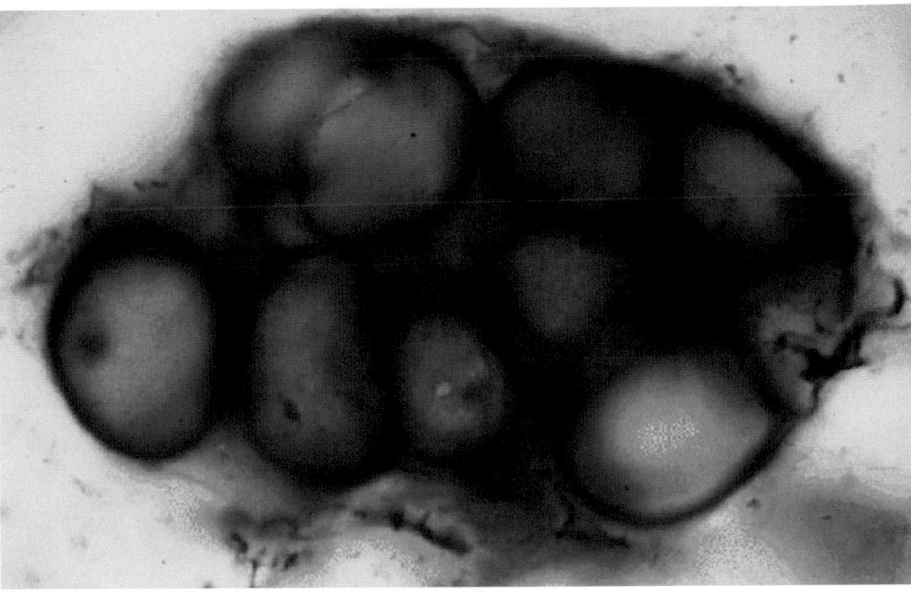

As has already been said, one of the biggest breakthroughs in the investigation of Precambrian life came in the 1950s – in Canadian strata. For some time, it was known that there were vast iron ore deposits in Precambrian rocks of great economic value, but little was known about their geology. Stanley Tyler's attention was drawn to a rock surface covered with small stromatolite mounds, and he recognized the rock as chert, a silica-rich marine deposit. Cherts are commonly associated with Precambrian ironstones and are commonly red in colour; however, here they were black and looked as if they had some curious, wispy lines within them, and so Tyler collected some specimens for analysis back in the laboratory at the University of Wisconsin.

With such fine-grained rocks, it is necessary to cut very thin slices so that they can be viewed with a high-powered microscope. When Tyler examined the slices, he was amazed to see that the wispy black lines were clearly organic filaments and were accompanied by lots of round, cell-shaped bodies. The position of the strata was deep within the Precambrian, and so Tyler knew that he was dealing with something that was remarkably old and of considerable interest. Being no kind of a palaeontologist, however, he needed to find someone who could tell him what they were.

Elso Barghoorn (1915–84) was a Harvard expert on fossil fungi and immediately recognized that Tyler had

Fossil bacteria

The 1965 discovery of the 850 million year old Bitter Springs microfossils from central Australia showed that microbial life was probably widespread throughout Precambrian times. This cyanobacteria-like colonial cluster of cells was recovered from laminated black cherts associated with stromatolites.

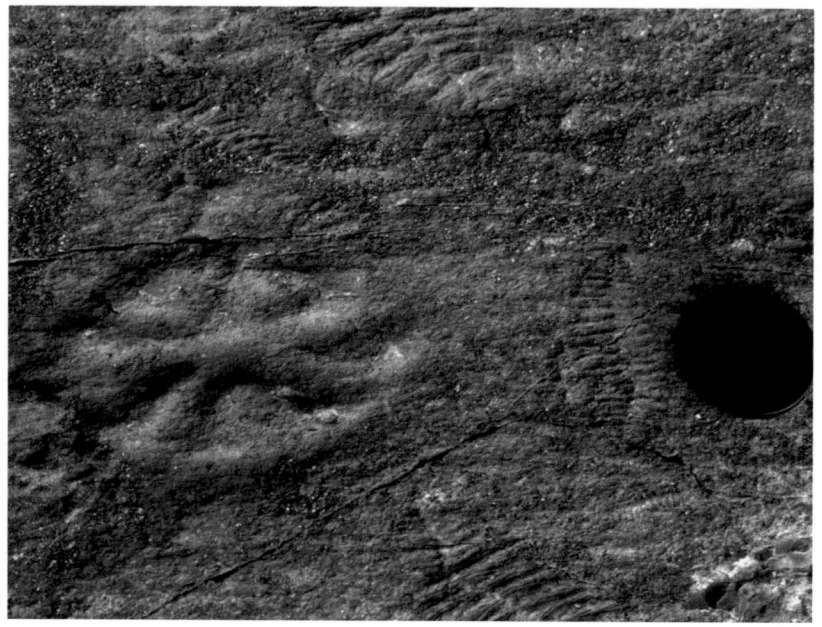

Mistaken Point Ediacarans
Ancient late-Precambrian seabed surfaces are exposed on the Atlantic cliffs of Cape Race in Newfoundland at a site named Mistaken Point. A variety of impressions of soft-bodied fossils have been found on these sea-bed sands, some of which can be closely matched with the contemporaneous Ediacaran fossil assemblages of Australia and elsewhere.

previously unmapped strata. With more than 4000 million years of Earth history to play with, most of it still unexplored, there is still enormous potential for making significant finds; however, for every one, there are many samples painstakingly collected in inhospitable terrain which turn out to be unproductive blanks. New fossils continue to be found and argued about. Traces of worm-like organisms can be difficult to verify since a number of inorganic processes produce worm-like marks in soft sediments. Still, at the other, younger, end of the Precambrian, another find made in the late 1940s has helped to revolutionize our view of life at this time — the Ediacarans.

The mysterious Ediacarans

The Ediacarans are perhaps the most intriguing group of fossils ever to have been found. In the 1960s, it gradually dawned on fossil experts that a unique group of fossils had been discovered in Australia more than 10 years previously, in the late 1940s. Entirely soft-bodied, they generally had a curiously quilted form and ranged from round jellyfish-like shapes, sometimes with radial symmetry or even tri-radial or bilateral, to some with "head" and other attachment discs. They were certainly difficult to shoehorn into known animal groups. Even the discovery of similar fossils in England did not immediately stir much interest. It was the publication of an article in *Scientific American* by Australian palaeontologist Martin Glaessner (1906–89) that probably set the ball rolling. There were intriguing and unresolved questions about these fossils: what kind of organisms were they, how and why were they preserved

stumbled upon something truly astonishing. Somehow or other, these tiny, soft cells, just defined by their cell membranes and unprotected by any shell or toughened organic coating, had become encapsulated in silica which had hardened quickly before they could rot away. This process had protected them for perhaps as much as 2,000 million years. Barghoorn knew that the find was so exceptional and so unlikely that he would have to be very careful and accurate in his description and illustration, otherwise the findings would not be believed. Indeed, when Tyler and Barghoorn first published their description of the Gunflint microbiota as including two kinds of fungi, two algae, and a unicellular protozoan animal, there were many experts who discarded the notion and thought that the samples must have somehow become contaminated with much younger material. Tyler and Barghoorn were right, however, and we now know that the Gunflint rocks are around 1,878 million years old.

Since this pioneering discovery, a whole new world of microscopic life in the seas of Precambrian times has been opened up by further finds from localities all around the world, but especially from Siberia, China, and Australia. Palaeontologists have learned from experience what kind of sedimentary environments and special circumstances tend to preserve these delicate microbial fossils and prospect for similar circumstances in

Finding the Ediacarans

Reginald Sprigg was an Australian mining geologist who was reinvestigating some abandoned gold workings in the Pound Quartzite of the Ediacaran Hills in South Australia when he found some strange jellyfish-like fossils. Sprigg was no paleontologist but in 1947 described the sandstone fossil moulds and casts as Cambrian in age. The find stimulated a Czech emigre, Martin Glaessner (1906-89), newly appointed as a paleontologist in the University of Adelaide, to have a closer look. Recognizing their potential significance, Glaessner, along with a colleague, Mary Wade, spent the rest of his working life on them.

as sandstone casts and moulds, and how did they fit into ideas about the evolution of life?

The early consensus of opinion was that they were soft-bodied creatures that were probably related to the jellyfish (scyphozoans), although there were some members of the biota that seemed to have a very unjellyfish-like bilateral symmetry and some that even seemed to have some kind of a head region, which was even more unjellyfish-like. Perhaps these latter forms represented early members of other major groups of organisms, such as the annelid worms and arthropods that turned up in the fossil record a few million years later when they had acquired fossilizable skeletons. If they were related to jellyfish, why were they so well preserved in sandstones which do not normally preserve such organisms? Was there something unusual about the environment of preservation at the time, or was there something unusually tough about the organic material of which the creatures were made?

Argument swung back and forth between those who thought the Ediacarans represented a unique group of organisms which subsequently became extinct and those who thought that many of them could be slotted into existing biological groups. If they had become extinct, did this mean that they did not have any evolutionary input to subsequent life? In which case, where in the fossil record were the forebears of the major living groups? Now, more than 50 years since they were first found in South Australia, a great deal more is known about the Ediacarans. Some questions have been answered, but just as many more have been raised.

The geological distribution of the Ediacarans through time is now much better constrained than previously. Sandwiched between the famous Cambrian "explosion" of life around 540 million years ago and the near global deep-freeze of the Varangian ice age, which ended around 565 million years ago, is the Vendian, when the Ediacarans lived. Fossil representatives of the biota have now been found on most continents and, while many are discrete to specific regions and marine environments, many others are cosmopolitan in their distribution. Some are common between very widely

Namibian Ediacarans
This long frond cast in sandstone from late Precambrian strata from Namibia is the Ediacaran fossil *Pteridinium* which probably lived partly buried in sea-bed sands and had a sack-like body which became filled with sediment.

separated localities, such as Newfoundland and Siberia, which at the time were located at different latitudes in different hemispheres. Such global distributions are very unusual, especially for organisms that lived either on or within the sea bed. As the world was still recovering from a near global ice age, however, it is possible that the world's oceans remained quite cold and that, as a consequence, climate zones were not yet re-established.

The Ediacarans show considerable diversity of form, ranging from simple round, flattened discs a few centimetres in diameter, through flattened frond and bush-shaped ones, to those which are more elongate with a distinct "head" end. Most are clearly visible, and the biggest are up to 2m (6½ft) in length, and as such they represent the first biota of largish organisms to have evolved or at least to have been preserved as fossils. Many but not all are preserved in sandstones and show some three-dimensional relief. The puzzle is that the form of many of their bodies is moulded both externally and internally with sand, as if they had open sack shapes. Some of these clearly lived within the sea bed and were able to grow and/or move upwards as more sediment was dumped on them. Detailed examination shows that some had elaborate quilted bodies made up of connected tubes which became filled with sand. This makes them certainly quite different from jellyfish, and many if not most were unlike any living creatures.

There are at least a dozen different groups of Ediacarans and altogether a hundred or so species known at present. The main groups include discoidal forms such as *Ediacaria* which were generally simple, flattened discs which do superficially look like jellyfish, but in detail show that they are different organisms and lived on the sea bed, often in dense populations. Another discoidal form represented by forms such as mawsonites had bodies made up of radially arranged sand-filled tubes or pouches. The vendobionts, such as *Pteridinium*

Stromatolite

By far the oldest visible fossils are laminated mounds of sediment known as stromatolites, some of which have been dated from as far back as 3,800 million years ago in early Precambrian times. The mounds range from a few centimetres in width and tens of centimetres high to a metre or so wide and several metres high.

Charnwood fossils

In the 1950s, Roger Mason, an English schoolboy with a passion for geology was looking over some patchy outcrops of ancient rocks in Charnwood Forest, near his home in Leicestershire when he found some strange fossil markings on a rock surface. The fossils are remarkably like those found in South Australia – casts and moulds of blob shapes, discs and fronds in coarse-grained sediments. They were described by Trevor Ford, a paleontologist at Leicester University, in 1958 as *Charnia masoni* and *Charniodiscus concentricus*. Since the strata above them can be matched with nearby Cambrian age strata, the Charnwood fossils also turn out to be late Precambrian in age.

and *Rangea* had sack-shaped bodies with a bilateral symmetry and clearly lived buried within the sands.

One of the best known of the Ediacaran groups is the dickinsoniids such as *Dickinsonia*, of which many examples are known. These grew from a small disc into an elongate, broad, flat ribbon shape up to 1m (3ft) long with a bilaterally dividing axis and thin segments or partitions on either side. These are slightly offset to one another, so that they are not true segments as seen in annelid worms. Perhaps related to these are the yorgiids (eg *Spriggina*) and vendomiids (eg *Vendomia*), which also had partitioned and disc- to more elongate-shaped bodies and a clear "polarity", in that one end was different from the other. There is also evidence that they were capable of creeping over the sediment surface, perhaps grazing on bacterial mats, but how they were able to do this without any teeth is not known.

Another well-known group grew as fronds (eg *Charnia*) and stood proud of the sea bed, attached to the sediment surface by round holdfasts. They were mostly up to 10cm (4in) high, but some have been discovered in Newfoundland which were 2m (6½ft) long. Occasionally, they were numerous enough to form tiny thickets covering the sea bed. Some experts have claimed that this group is related to the living seapens (pennatulaceans), which they do indeed resemble. There is also evidence, however, that the upright "pen" of the Ediacarans was stiff and often became buried in an upright position by sand, which is quite unlike the soft-bodied seapens. A further variation on the frond theme is a multi-branched frond (eg *Bradgatia*), which may have grown out across the sediment surface.

One of the most remarkable of the Ediacaran groups is that of the trilobozoans (eg *Tribrachidium*), which have a tri-radially symmetrical body which is most unusual. Again, they are disc-shaped, and well-preserved fossils from the White Sea region of Arctic Russia preserve details never seen before, showing that they are quite unlike any living forms. There are a number of other groups, including trace fossils, which are mostly surface trails that must have been made by elongate, worm-like bilaterally symmetrical organisms grazing the sediment surface for microbial food. Others seem to have lived in vertical burrows from which they "vacuumed" or grazed the surrounding surface, leaving patterns of semi-radial scratches.

Perhaps one of the least spectacular but most important elements of the biota are the microbial mats which covered many of the sediment surfaces and are preserved as carbonaceous sheets in fine-grained sediment layers, such as preserved in deep boreholes from the White Sea region. Mud rocks recovered in drill cores from as deep as 3.5km (2 miles) are remarkably well preserved, with no deformation apart from flattening. Within these layers, round carbonized discs have been found which are probably reproductive cysts of algae forming part of the microbial mats. Evidently these mats formed the base of the food chain for at least some of the Ediacarans, although others might have been filterers that fed on micro-organisms within the seawater. Almost all the Ediacarans appear to have bodies made up of thin sheets of tissue that grew into various flat discs and tube shapes which, in the case of those that lived within the sediment, were filled with sand.

Much is still to be discovered about the Ediacarans, especially as to whether they gave rise to any surviving groups such as the seapens (pennatulaceans). Certainly there are some very close resemblances between living seapens and some of the Ediacarans. Furthermore, there are Cambrian seapen-like fossils from the Burgess Shale that make the connection more likely. Another interesting development has been the discovery of fossils of some of the other Ediacarans surviving through into late Cambrian times. Assigned to the genera *Ediacaria* and *Nimbia* they were found in some relatively deep-water turbiditic deposits in the south of Ireland.

Discovering the origin of life

Charles Darwin speculated in his groundbreaking book *On the Origin of Species by Means of Natural Selection* that "all the organic beings which have ever lived on this Earth may be descended from some one primordial form". And while late 19th-century geologists were busy finding more and more Precambrian strata, but still no really good fossils, biologists were uncovering an ever increasing and astonishing diversity of life on Earth, especially water-living primitive micro-organisms. With Darwin's theory of evolution pretty well established, biologists soon realized that there must have been a number of evolutionary stages as life became more complex in its organization. It also became evident that the diversity of living organisms retains a good sample of those early stages. These range from primitive unicells through multicellular organisms (metazoans) such as simple sponge-like creatures through those with two-cell layers, such as jellyfish, to three-cell layers, such as flat worms, to those with an internal space known as a "coelom", such as round worms etc.

Consider for a moment what some of these evolutionary stages might be. Compare the biology of birds or flowers with that of sponges or red algae (such as a simple seaweed). Birds and flowering plants are much more complex than sponges or seaweed. They have more structured bodies with organs, transport systems, and different tissue types, all made up of many cell types. By contrast, neither sponges nor red algae have organs or transport systems. Differences between their cells are minimal, but these life forms are still made up of many cells. Single-celled organisms show an even more primitive condition in which there is no differentiation of cell types. All body functions have to be contained within the single cell.

Assuming that life went through these evolutionary stages, the fossil record should reflect this. We should find fossils of relatively primitive organisms such as sponges and red algae before those of more complex organisms such as birds and flowers. The temporal distribution of such fossils was already evident by the middle of the 19th century. Additionally, however, we should find fossils of single-celled organisms before those of sponges and red algae. Many marine sponges have fossilizable hard parts and their petrified remains indeed have been found within Cambrian rocks since Sedgwick's day, but so have the fossils of more complex organisms such as trilobites. So it seemed highly likely that there would have been primitive life in the Precambrian, but the chances of it being preserved seemed very remote at the time. This view was eventually confounded, but, in the meantime, some biochemists tried a new approach to the question of the origin of life.

The biochemical broth of life

One other approach to the intriguing problem of early life is experimental. Organic chemistry was founded in the early 19th century, when Friedrich Wohler (1800–82) first synthesized urea in the laboratory. Since then, it has become clear that organic chemicals can be generated without the intervention of living organisms. It was to take another century, however, before anyone thought that organic molecules might have formed in early oceans by spontaneous non-biological processes and then become capable of self-replication. In 1871, the Cambridge botanist Joseph Hooker (1817–1911), who had been one of Darwin's main mentors when he was a student, received a letter from Darwin commenting: "It is often said that all of the conditions for the first production of a living organism are now present, which could ever have been present. But if (and oh! what a big if!) we could conceive in some warm pond, with all sorts of ammonia and phosphoric salts, lights, heat, electricity etc. present, that a protein compound was chemically formed ready to undergo still more complex changes, at the present day such matter would be instantly devoured or absorbed, which would not have been the case before living creatures were formed."

It was not until the 1920s that anyone seriously took up Darwin's challenge of replication "creation" with

a bit of chemosynthesis in the laboratory. Alexander Ivanovich Oparin (1894–1980), a young Russian biochemist, noticed that dispersed oil forms droplets which superficially resemble simple spherical cells. He speculated that if the four basic and abundant elements – carbon, hydrogen, nitrogen, and oxygen – combined together, they could form the organic "elements" of life. Small and persistent molecules such as carbon dioxide (CO_2), ammonia (NH_3) and methane (CH_4) could perhaps then combine in water to form more complex and bigger molecules. With time, this imaginary "primordial soup" might provide the milieu in which there could be further synthesis, perhaps leading to the formation of enzymes and, finally, complex proteins – genes which might be self-replicating (not that Oparin or anyone else had much of an idea what genes really were at that date). Oparin's ideas were initially published in Russian and were not communicated to the West until 1938, when an English translation became

available. The result was that there was another delay before scientists were to have a further attempt at solving the problem.

The 1950s saw the beginning of a renaissance in the investigation of life's biological origins. It was an American physical chemist, Harold Urey (1893–1981), who became interested in the question and thought that the primitive Earth probably had a reducing atmosphere. He encouraged one of his students and later co-worker chemist, Stanley Miller (b.1930), to try a series of lab experiments based on this notion. In 1953, Miller passed a mixture of methane, ammonia, and hydrogen through water into which electrical energy was continuously sparked. Miller thought that this replicated in the simplest possible way the biochemical and physical conditions of early Earth. Within days, an impressive array of organic molecules were indeed synthesized, including 25 amino acids, which are the building blocks of proteins. Since then, other energy sources have been

Modern stromatolites

Stromatolites are among the oldest biological structures preserved in the rock record. Luckily we can get a good idea of how they formed in the Precambrian past because they still survive in some shallow tropical seas, such as Shark Bay, Western Australia. Microbial mats grow over the sea bed in patches which get periodically covered with sediment. In order to survive, the microbial filaments grow up through the sediment to form a new surface mat and in so doing incorporate the layer of grains, gradually building mound-shaped structures.

tried, including radiation and ultraviolet light, and even more complex molecules have been synthesized, including adenine, the base for nucleic acid. But how such molecules might have evolved into life is not clear. What would organize them? There is still no obvious mechanism – yet.

Meanwhile, this line of research was eclipsed by a much more important discovery, that of the structure of DNA by Francis Crick and Jim Watson in 1953. The double-helix structure of the nucleic acid DNA is now seen as the key to the origin of life, with DNA making RNA making protein. In 1965, Nobel Laureate Linus Pauling (1901–94) and Emile Zuckerkandl (b. 1922) suggested that evolutionary history can be read in biological molecules. A decade later, Carl Woese and George Fox of the University of Illinois made a comparison of RNA sequences which does just that. They produced a "tree of life" which would have delighted Darwin.

Three great groupings were recognized: the Eucarya, which include animals, plants, fungi, slime, moulds etc; the Bacteria which are a much bigger group than generally appreciated; and a third large group, the Archaea, which are prokaryote microbes and generally unfamiliar to all but experts in the field. The Archaea include microbes that can live in extreme conditions, many of which do not need oxygen to survive, but derive their energy from chemical reactions between sulphur and hydrogen compounds.

In 2001, it was discovered that archaeans may be the most abundant organisms in parts of the oceans. Interest has also focused on the extremophile members of the group which can tolerate high salinities (10 times saltier than the oceans), temperatures (up to 113°C (235°F)), pressures, and acidity (pH1), and others at the opposite end of the spectrum which can tolerate very low temperatures etc. Such microbes would have prospered during the early evolution of the Earth, and a better understanding of them, especially their genetics, is likely to provide new insights into the origin of life. But it is not going to be easy.

It has already been revealed that horizontal transfer of genes does occur, especially among these so-called primitive organisms. We have not got to the root of the "tree of life" yet, and some experts doubt whether we ever will. The simplistic idea that a thin gruel of amino acids might be sparked into life is clearly naive. Chemosynthesis may well have played its part, and we now know that it is much more common on Earth than previously suspected. It does happen in the depths of the oceans around sea floor spreading ridges with hydrothermal vents, and deep underground, and it may have happened in space. The discovery that some meteorites contain carbonaceous molecules opened up a new possibility.

Meteorites are a diverse bunch of rocks derived from space that are a very important source of information about the age of our solar system and the early stages in development of the Earth. One group is known as the chondrites, some of which originated around 4,560 million ago and are the oldest survivors of the birth of the solar system. Many chondrites have also been found to contain organic compounds, especially one group that is generally known as carbonaceous chondrites. Some early analyses of these did not take account of subsequent contamination by Earth-derived organic materials, but more recent analyses record a remarkable story.

Complex mixtures of straight hydrocarbon chains (alkanes), rings of aromatic hydrocarbons, amino acids, and recently some sugars (saccharides) have been found in these meteorites. However, they all originated from much simpler molecules which were originally formed in interstellar space. The synthesizing process went on within solar nebula and the planetismals which bombarded Earth during its early formation. It is now thought that meteorites brought much of the initial organic building blocks for life on Earth during that early bombardment. But that still leaves the big unanswered question: how did such organic molecules acquire the "spark of life"?

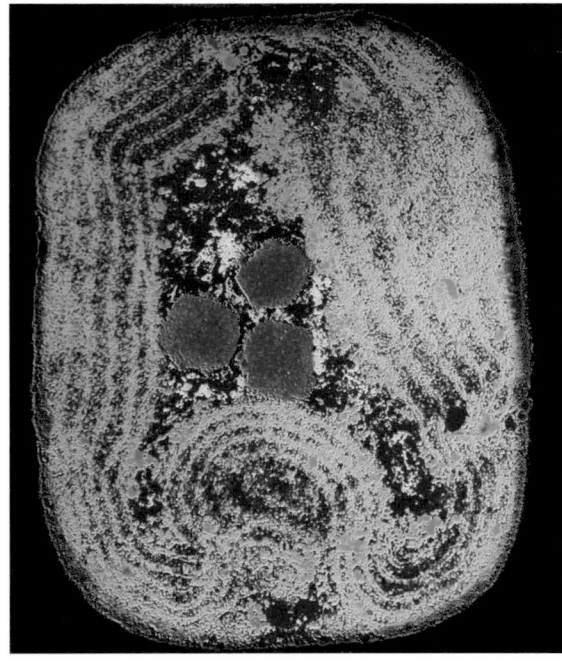

A prokaryote

This cell from a filamentous cyanobacterium typifies the predominant micro-organisms which form stromatolites, and has the primitive structure of a prokaryote. Its nuclear material is not surrounded or contained by a nuclear membrane, as found in all more complex eukaryotic organisms, and the concentric inner folds are photosynthetic membranes.

Methane seep

Methane gas (CH4) is one of the basic organic compounds to have been synthesized during the early formation of the Earth, and certain bacteria can tolerate living in methane-rich environments. The methane seep seen here from Taiwan is in a fault zone in a non-volcanic region at Kuantzuling on the western side of the island. The methane is probably derived from buried black shales and has been lit either by lightning or human intervention.

The trouble with fossils

You might think that, after several hundred years of scientific investigation of fossils, paleontologists would be able to put hand on heart and say without equivocation whether something is or is not a fossil. The distinction can still present an interesting problem, however, and one which is not just of historical interest. Witness all the controversy over the putative Martian fossils, claimed to have been found in a meteorite.

Space is full of carbonaceous material left over from the early formation of the universe, and there should therefore be many, many other planets with environments just as hospitable to life as the Earth's. This knowledge has finally overcome the scientific and intellectual barriers not only to realizing that there might be life elsewhere, but also to considering that there ought to be. Even within our solar system, our

understanding of extremophile organisms now raises the possibility that there could be primitive life on Mars despite its incredibly inhospitable environments. One way to check this is to go there and look, but this option is expensive. For some time, it has been known that among the collection of meteorites found on Earth are some which have probably come from Mars. Consequently, it dawned on space scientists that careful examination of them might resolve the problem — perhaps fossil microbes or chemical signs of life could be preserved in such a meteorite.

Microscopic examination revealed some promising signs and, in August 1996, headlines declaring "PAST LIFE FOUND ON MARS" reverberated around the world. There were complex carbon molecules, tiny magnetite (an iron oxide mineral) grains, and a microbe-like tubule within

Mars meteorite

This is the meteorite (known by its catalogue number ALH84001) which is believed to contain fossil evidence for life on Mars. Found in Antarctica in 1984, the rock has been dated at about 4500 million years old, ie it is about the same age as the Earth. The rock was blown off Mars by a huge meteorite impact about 16 million years ago before falling to Earth about 16,000 years ago in the Alan Hills ice field, Antarctica, where it was found.

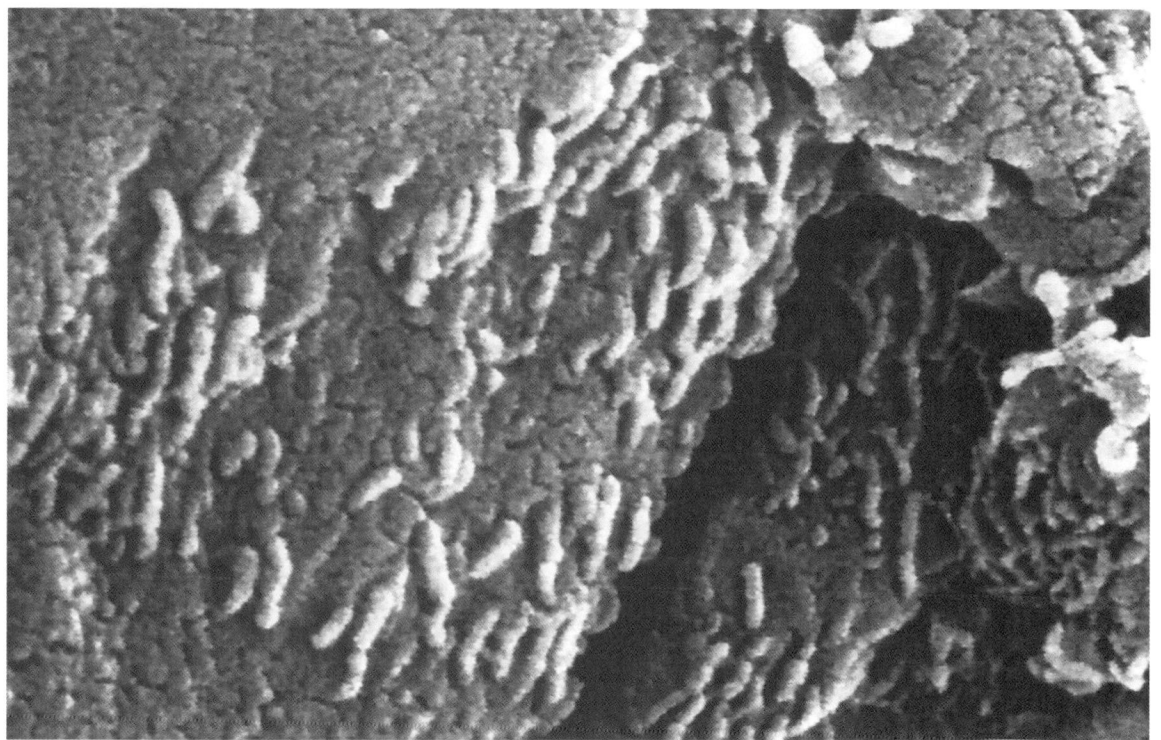

Life on Mars?
In 1996, newspaper headlines around the world were announcing the discovery of evidence for life on Mars and publishing photos of tiny rod-shaped structures which were claimed to be bacteria. The rock containing them was a meteorite, found on Earth but derived from Mars. More detailed analysis showed that although the "bugs" looked organic they are more likely to be inorganic structures.

carbonate minerals from the meteorite. Closer examination and analysis, however, raised doubts about their organic origin. It turns out that the complex structured hydrocarbons have Earth-derived radiocarbon isotopes and so are contaminants, rather than derived from Mars. The tubule was thought to be a small fossil bacterial rod, but is an order of magnitude smaller than any similar living structure. The magnetites were compared with intracellular magnetite grains found in some particular kinds of terrestrial bacteria. Their detailed structure was shown to have resulted from mineral formation at high temperature and therefore unlikely to have occurred within bacterial cells. This does not, however, discount the possibility of life on Mars, nor is there any doubt that extraterrestrial hydrocarbons are preserved in certain carbon-rich meteorites. But it does show that the question of what is and what is not a fossil is still very much alive and of some importance. Another example of just how remarkable fossil preservation can be is provided by the Doushanto babies.

Frozen in time on a sea bed around 570 million years ago, clusters of tiny balls, less than a tenth of a millimetre in diameter, might not at first seem very

spectacular or interesting, but they promise to revolutionize our ideas about the early evolution of animals and plants. They were discovered in the 1980s, buried in rock strata of the Doushanto rock formation, southern China. Never before have fossils of such delicate structures been found. To the astonishment of the scientific world, these minute balls were described (in 1998) as the fossilized embryos of primitive animals and plants. Further, they present supporting palaeontological evidence for the divergence of animal phyla in the "deep" Precambrian past, long before the start of the main fossil record at the beginning of Cambrian times, around 544 million years ago. The Chinese fossil embryos are one of the most exciting palaeontological finds of the end of the 20th century. They open a window on a phase in the history of life which is otherwise very poorly represented by fossils.

There is absolutely no doubt that these are genuine embryos, fossilized in the very early stages of cell division (cleavage) following fertilization. Electron microscope images of the fossils show clusters of cells from the first two-cell division through to blackberry-like spheres and cube-shaped packages of cells. They are all beautifully preserved in phosphatic minerals. And their form is identical to the different stages in cell division of living organisms. It appears that, within hours of being shed into the sea by their parents, these eggs were fertilized, killed, and preserved so well that they have survived for 570 million years.

Normally, soft cell material is not preserved by fossilization; however, in exceptional conditions, tough tissue, such as muscle, may resist putrefaction long enough to be dehydrated (such as in amber) or coated by bacterially released minerals and thus replicated (as in this case) and preserved. But cell material is so "watery" and the bounding cell membrane so delicate that it normally ruptures within hours of death and is destroyed. In this instance, the process of fossilization must have been extremely rapid, with entombment in anoxic, phosphate-rich water and mud which was full of bacteria. The rapid reproduction of the bacteria changed the chemistry of the watery mud and led to the precipitation of phosphate mineral coatings to the cells before the membranes broke down.

Unfortunately, the patterns of early cell division in embryos are not particularly diagnostic for individual groups of animals. Some configurations of these fossil embryos can be identified as typical of algae and sponges, however, and the latter is confirmed by the presence of sponge spicules in surrounding sediments. More important are those embryos which are thought to be characteristic of flatworms, nematodes, and arthropods. This group of invertebrate animals is particularly interesting because they are multicellular organisms (metazoans) with a more advanced structure than sponges or algae. If the identification is correct, it implies that complex metazoans must have evolved well before 570 million years ago and are just not represented in the fossil record – or at least have not been found yet.

The discovery and identification of the Doushanto embryos lends support to an estimate for the timing of metazoan evolution as suggested by the so-called "molecular clock". This clock is calibrated to an estimated standard rate of genetic change. By measuring the genetic "distance" (comparison of gene sequences) between groups of living organisms, the time of original divergence can be calculated for a particular estimate of rate of genetic change. Recent attempts to do this have

produced divergence dates for the metazoans as far back as 1.2 billion years, but other modifications of this molecular-clock method suggest a more reasonable figure of 670 million years. This latter figure is more generally accepted by palaeontologists, and the Chinese embryos seem to support this more conservative estimate.

The unique role of phosphate in the preservation and fossilization of soft tissue is now much better understood and appreciated by palaeontologists. They are now actively prospecting for other sedimentary phosphate deposits which might provide similarly preserved and privileged windows on the past. The potential is enormous both within the Precambrian and younger Phanerozoic strata.

The fossil record normally preserves just the hard parts of organisms, the shells, teeth, bones, and wood, which are pretty tough. The remains have to be robust enough to survive being washed around by river or seawater and being buried in mud or sand. They then have to last through millions of years while the sediment is transformed by compression and chemical change into rock. Soft body tissues are not normally preserved, so all those organisms which have soft bodies are under-represented in the fossil record, and many are not found at all, especially those that are small. This means that there is no fossil record of the early embryonic stages of life, when all multicellular organisms consist of incredibly small balls of dividing cells. At least, this was the case until a decade or so ago when the revolutionary Chinese fossil embryos were found.

Before this remarkable discovery, evidence for the evolution of complex organisms this far back in time was highly contentious. While primitive life, as recorded by chemical fossils, began as far back as 3,800 million years ago, it took an astonishing length of time – about 2,500 million years – before the evolution of sexually reproducing single-celled organisms. Multicellular algae turn up in the fossil record around the same time, some 1,200 million years ago, but there was no clear fossil evidence for many-celled (metazoan) animals. Of course, there are the puzzling Ediacaran fossils, and it is virtually impossible to conceive of such large organisms (up to 2m (6½ft) in length) being unicells, assuming that they are metazoans.

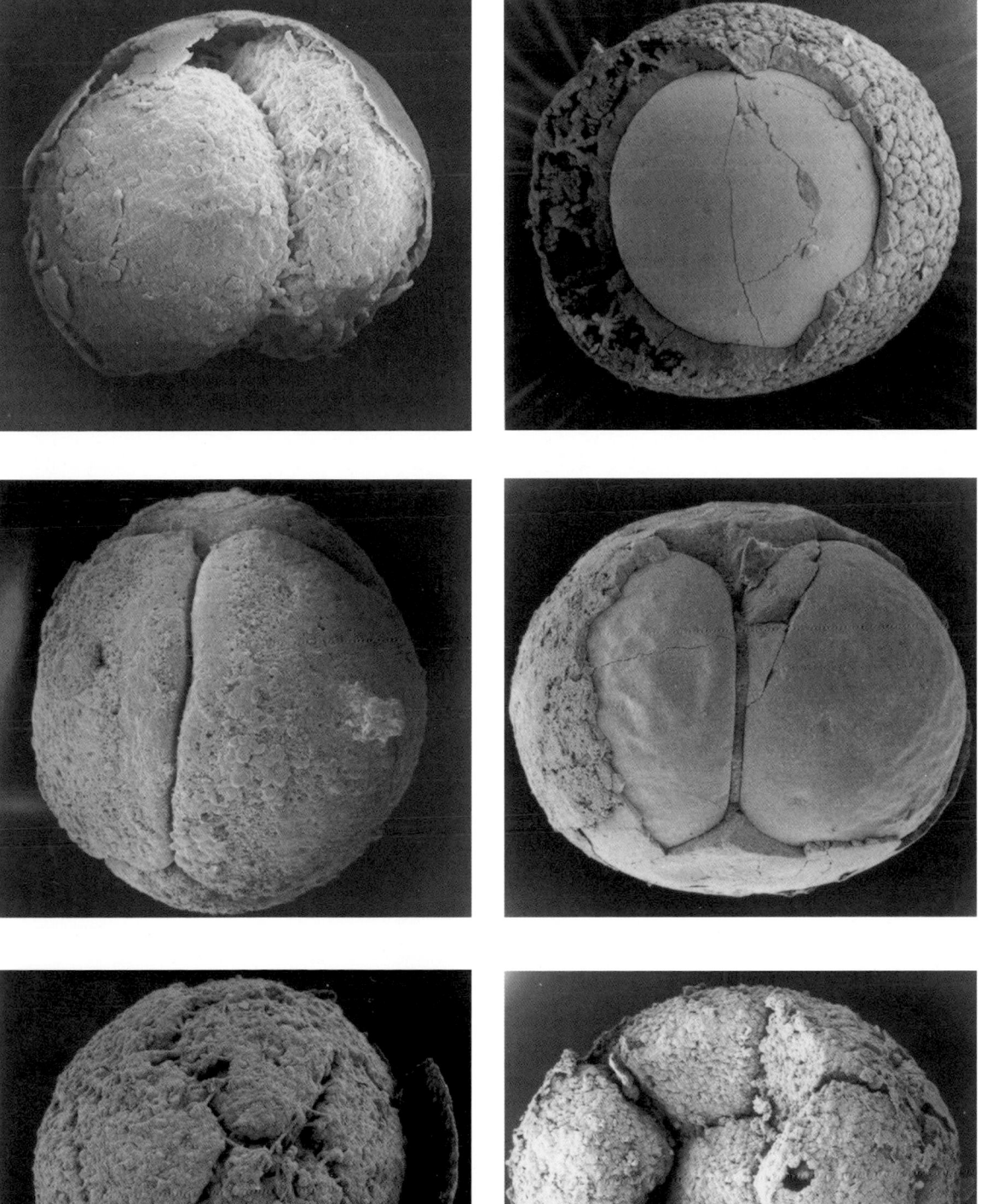

Doushanto embryos
These amazing microscopic spheres are embryos of sea-dwelling invertebrate animals and show successive cleavage stages in cell division. Their parents lived in late Precambrian times (some 570 million years ago) and the embryos were found in phosphate-rich sediments in southern China.

8
Earth's beginnings

Speculation about Earth's beginnings is probably as ancient as the gift of speech – and that may well predate our species, *Homo sapiens* – but it is unlikely that we will ever know the whole answer. Certainly the first signs that can be construed as astronomical speculation are associated with rock art generated by our kind some tens of thousands of years ago. Since then observers from many different cultures over the last few millennia have gone to great lengths to calculate the movements of planetary bodies and have achieved remarkable insights about the position of Earth in relation to the Sun and the other planets. However, it is only since the invention of the astronomical telescope and the advent of intellectual freedom to speculate about Earth's origins unfettered by religous considerations that modern understanding has developed, especially over the last 200 years or so.

PRECAMBRIAN EON

| ERA | 4600 MA HADEAN | 3800 MA ARCHEAN |

HADEAN ERA

periods not yet defined | 4600 MA |

4500 MA Impact dislodges moon-sized mass from the Earth

4400 MA Oldest minerals (zircons); differentiation of Earth's core

4200 MA Atmosphere and oceans formed

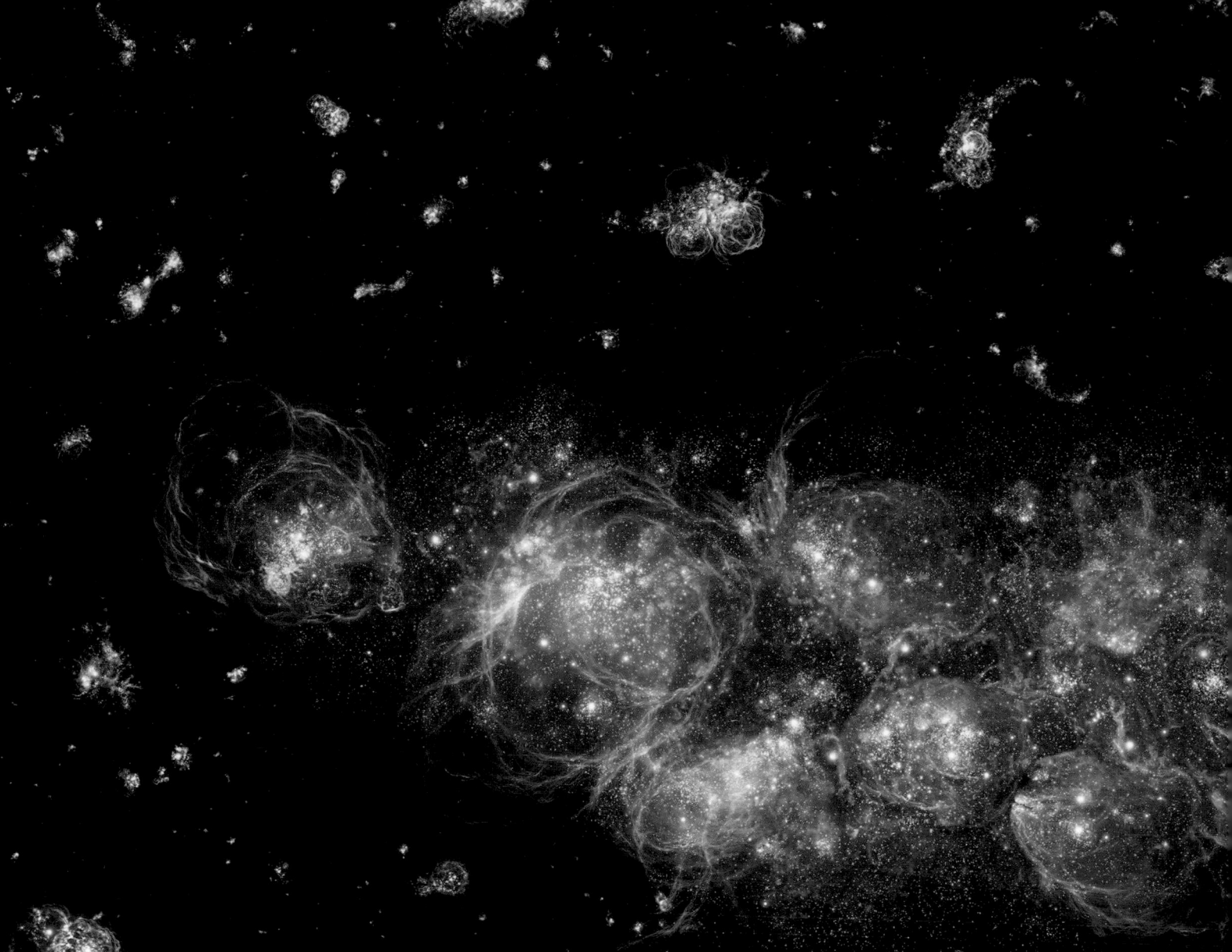

The contribution from the rock materials of the Earth to the debate about its origins is much more recent. The dynamic nature of the Earth has meant that the first formed rocks on Earth have either been eroded away or deeply buried and are thus unavailable for sampling. Even those that have been brought to the surface by subsequent earth movements have mostly been highly deformed and altered from their original state by heat and pressure, a process know as metamorphism, which also tends to destroy any fossil remains within them.

Much has depended upon the ability to date rock materials and to know where to look for the oldest rocks on Earth. Although the principles of such dating have more or less been known about since the discovery of radioactivity at the end of the 19th century, the techniques for doing so took several decades to develop. It was only in the wake of the Manhattan Project to develop the atomic bomb that radiometric dating began to be reliable and the age of the Earth was finally "pinned down" to around 4,550 million years.

Early universe
This 1,000-million-year-old universe, with stars being born, is based on deep field observations by the Hubble Space telescope. Because light has taken billions of years to reach Earth from these distances, these observations look back to the early history of the universe.

1200 MA The first multi-celled organisms date from the middle of the Proterozoic period

610 MA The first large marine animals appear

PHANEROZOIC EON

2500 MA PROTEROZOIC

545 MA PALEOZOIC | 248 MA MESOZOIC | 65 MA CENOZOIC

3900 MA Bombardment destroys early atmosphere and oceans

3800 MA first evidence of chemical life on the planet

Discovering the age of the earth

Noah's Ark
The Judeo-Christian tradition and culture is founded on the Old Testament of the Bible, with its accounts of the Creation, the Fall of Man and the Flood. This was sent by God to drown all the sinners apart from Noah, his family and a sample of all life which was to restock the Earth once the Flood had subsided. Flood stories exist in many cultures, especially in the Middle East, because of extensive flooding from the meltwaters of Ice Age glaciers and ice sheets.

The scientific investigation of Earth's beginnings – how and when it was formed and what might have preceded it – has really only begun to get under way during the past 200 years. For much of recorded history, ideas about Earth's age and origins were based on religious doctrine. In 17th-century Europe, almost everyone believed that the world was created in six days about 6,000 years ago and that the biblical book of Genesis contained an accurate description of events. Some biblical scholars believed that even greater precision was possible. For example, in 1642, John Lightfoot (1602–75), a vice chancellor of Cambridge University, stated that the world was created at 9 am on 17 September 3928 BC, a calculation based on studying biblical genealogies.

Not everyone believed in a 6000-year-old Earth. One investigator who was convinced it must be older was a Frenchman, the Comte de Buffon (1707–88). In 1760, Buffon carried out the first estimate of the Earth's age based on a reasoned scientific approach and experiments. He postulated that Earth must have started out as a very hot, molten sphere and had since cooled to its present state. To estimate how long that had taken, Buffon experimented with heating iron spheres and scaling their cooling to an Earth-sized mass. The answer he obtained for the Earth's age was 75,000 years. A few years later, after pressure from Church authorities, he was forced to retract this claim.

Despite this setback, by the early 1800s, belief in a 6,000-year-old Earth was crumbling. There was a growing support for the doctrine of "uniformitarianism", which argued that Earth's features were formed primarily by slow, gradual changes over time, of a type still occurring today. It was becoming clear, for example, that immensely long periods of erosion had shaped the planet everywhere. The idea of uniformitarianism was first introduced in 1795 by a Scottish scientist, James Hutton (1726–97), in his book *The Theory of the Earth*. His views contrasted sharply with an opposing doctrine, catastrophism, which proposed that a series of great catastrophes (such as the biblical Flood) were responsible for most of the features visible on Earth's surface today.

One process that uniformitarians studied closely was the formation of sedimentary rocks. By the 19th century, geologists already had a fairly complete picture of the order in which the world's sedimentary rock strata had been lain down throughout geological time. They were able to obtain rough estimates of the total thicknesses of these sediments, ranging from about 25,000–112,000m (82,000–367,500ft). They also observed present-day sedimentation rates (in the order of a few centimetres per 100 years). Assuming these had remained roughly the same throughout geological time, it was argued that Earth's age could be estimated by dividing the total thickness of sediments by the sedimentation rate.

One such estimate, carried out in 1860 by the English geologist John Phillips (1800–74), gave a figure

of 96 million years, but this was just one of many 19th-century estimates using the same methodology, which gave ages ranging from three million to more than a billion years. There were a number of problems with such efforts – the sedimentary record was still being constructed and, because the sedimentation rate has undoubtedly varied throughout time, any figure used for it could only be a rough approximation. Nevertheless, people's ideas about the Earth's age were changing.

Two of the most important 19th-century proponents of an ancient Earth were Sir Charles Lyell and Charles Darwin. Lyell was a leading proponent of uniformitarianism. He was convinced from an early age

that the Earth was extremely old and set out to find evidence to prove it. Between 1830 and 1833, Lyell published his findings in a highly influential book, *Principles of Geology*. Charles Darwin took a copy of the book on his five-year voyage around the world in the *Beagle* (1831–36) and later applied Lyell's notion of gradual change to his theory of evolution by natural selection. As Darwin developed his theory, he realized that it must have taken hundreds of millions of years for natural selection to produce the astonishing range of species that exist on Earth today. He, too, became convinced of Earth's great antiquity. In 1859, in the first edition of *On the Origin of Species*, Darwin made a crude calculation of how long it might have taken for erosion

Noah's family tree

The Old Testament contained complex accounts and lists of genealogies from which early scholars tried to work out the duration of time that had elapsed since the Creation. Numerous such attempts were made, several of which arrived at dates of around 4000 years before the birth of Jesus. Belief in such chronologies was widespread even in the early part of the 19th century.

to wash away the Weald, a notable valley area in southeast England. He obtained a number for the "denudation of the Weald" (which roughly corresponds to the period of time since the end of the Cretaceous period) of 300 million years. Darwin was satisfied that this was the type of time scale required to fit in with his theories – although we know now that his figure was a considerable overestimate.

In 1862, a new figure was to enter the fray. This was the eminent Scottish physicist William Thomson (1824–1907), better known as Lord Kelvin, who was sceptical of Darwin's theory of evolution. Like Buffon before him, Kelvin asked the question: how long has it taken for the Earth to cool from its initial heat of formation? To answer this, he relied on a new theory of heat conduction developed by the Frenchman Joseph Fourier (1768–1830), which used some advanced mathematics. Kelvin estimated that Earth had begun as molten rock at a temperature of 3871°C (7000°F), and he knew that its temperature increased by 0.55°C (1°F)

for each 15m (49ft) descent into the ground. Using Fourier's theory, he calculated that it had taken 98 million years for the Earth to cool to this state. This was a minimum acceptable age consistent with Lyell's theory of gradual geological change; however, it was too short a period of time to accord with Darwin's theory of evolution. Nevertheless, Kelvin's calculations had to be taken seriously because he was such an authoritative figure. Later, in 1897, he revised his estimate downward to 20–40 million years. This was too short for both the geologists and the evolutionary biologists.

We know today that Kelvin's calculations were wrong because of an error in his initial assumptions. Kelvin assumed that Earth has no internal source of heat, whereas in fact the phenomenon of radioactivity provides a significant and constant source of heat throughout the Earth's interior. This is the main reason that Kelvin's estimate of the Earth's age was far too low. In the end, it was the later French discovery of radioactivity that would lead to a revolutionary new

Zircon crystal

Some minerals such as zircon contain radioactive elements which "decay" at known rates over time to form a succession of so-called "daughter" isotopes. By measuring the remaining proportion of isotopes and knowing the rate of decay, it is possible to calculate the date at which the mineral was first formed and, if it was formed in an igneous rock from molten magma, the date of the rock's formation. A number of measurements have been made on this zircon crystal, from which a mean age can be calculated within a certain margin of error.

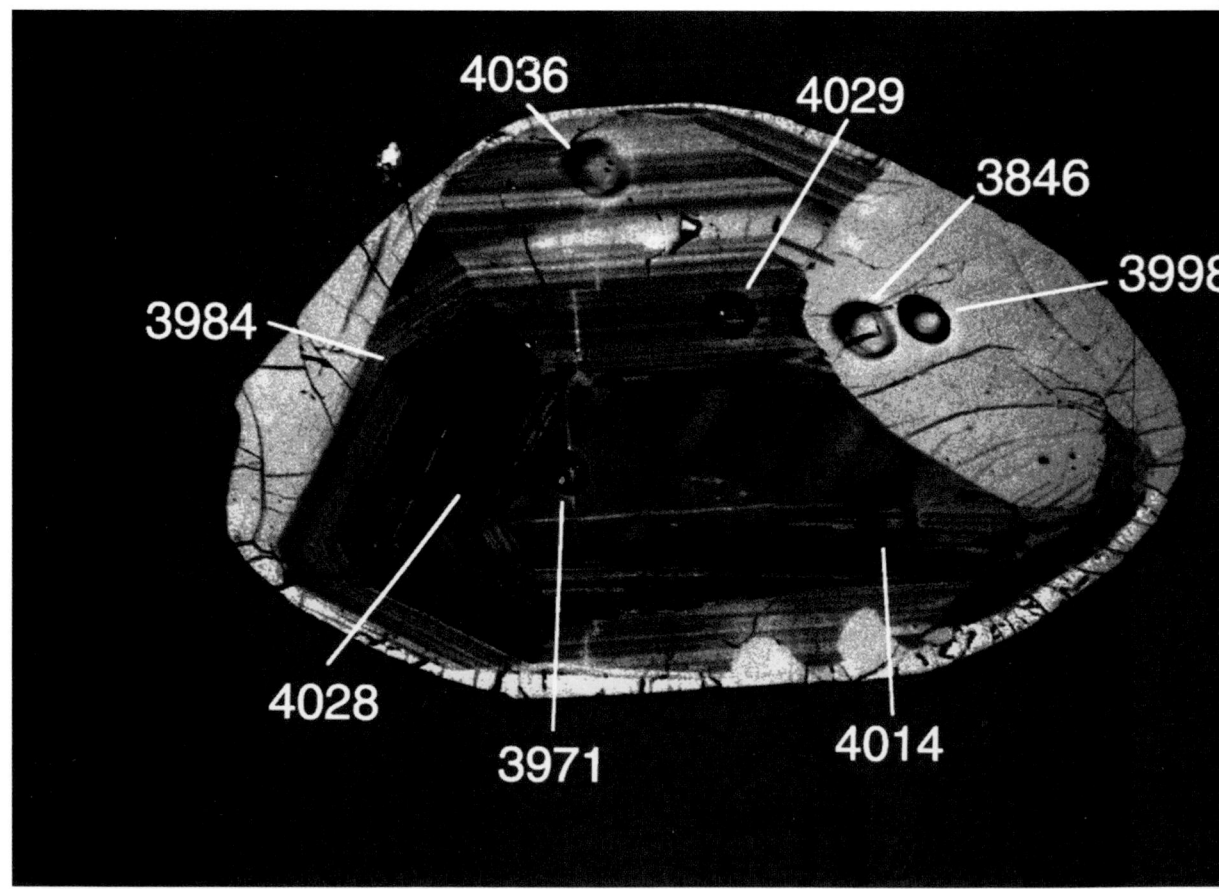

method for dating the Earth and its rocks. The method it led to was called "radiometric dating".

Radiometric dating

The origins of radiometric dating go back to 1896, when the French physicist Henri Becquerel (1852–1908) announced that the salts of uranium emit a type of invisible "radiation" that could fog photographic plates. A young Polish scientist, Marie Curie (1867–1934), and her husband, Pierre (1859–1906), took up the study of these mysterious "Becquerel rays". They concluded that the radiation was an inherent property of uranium atoms, which they termed "radioactivity". Within two years, they had shown that another element, thorium, is also radioactive and had discovered two new, highly radioactive elements, radium and polonium.

Over the next few years, many further discoveries were made about radioactivity, most of them by Ernest Rutherford (1871–1937), a brilliant New Zealand-born physicist. Working with British chemist Frederick Soddy (1877–1956), Rutherford demonstrated that, when the atoms of a radioactive element give off what they called an "alpha" or "beta" ray, a new element is produced – a phenomenon they called "decay". Furthermore, the decay product was often itself radioactive and could change over time into a third element and on through a whole series of transformations.

Rutherford established that a specific radioactive element decays at a precise rate, unaffected by external variables such as temperature or pressure. He coined the term "half-life" to quantify an element's rate of decay. This is the amount of time needed for half a sample of a substance to change into its decay product. Uranium and thorium seemed quite stable, with half-lives measured in billions of years, but other elements had much shorter half-lives, some as little as a fraction of a second. It occurred to Rutherford that radioactivity might be used as a "clock" to measure the ages of rocks and minerals. When a rock is formed, any uranium atoms within its structure would be "locked in". Over millions of years, the uranium atoms would gradually change into other atoms in a decay series. By comparing the amount of uranium left in a rock to the concentrations of its decay products, it should be possible to ascertain the rock's age.

An American chemist, Bertram Boltwood (1870–1927), took up Rutherford's idea and studied the chemistry of uranium and thorium-containing minerals. He soon established that both of these radioactive elements decay through a series of intermediates to a final stable decay product – atoms of lead, which are not radioactive. After some mathematical analysis, Boltwood realized that he had a viable radiometric "clock". The age of a mineral could be approximated by its ratio of lead to uranium (Pb/U) divided by the rate of uranium decay. Armed with a Geiger counter to measure decay rates and some chemical apparatus to detect lead and uranium concentrations, Boltwood began testing minerals from rocks whose relative ages were already known. In 1907, he published his calculated ages for 10 mineral occurrences. These ranged from 410 million years for a uraninite (uranium oxide) from Glastonbury, Connecticut, to 2,200 million years for a thorianite (thorium and uranium oxide) from Ceylon (now Sri Lanka). These figures implied a dramatically more ancient Earth than anyone had suggested before.

The importance of radiometric dating to understanding the deep history of the Earth is inestimable. Previously, geologists could only be sure of the relative ages of rocks through using stratigraphy (the geological column). Radiometric dating, however, gives absolute ages. This offered the possibility not only of knowing the age of particular rocks, but also of finding out when particular geological eras and periods began and ended, how long they lasted, and how old the Earth really was. Estimates of Earth's antiquity had been made before, but it was clear to many scientists that they were just guesstimates, usually based on questionable assumptions and incomplete knowledge. Radiometric dating seemed more solid – it was based on what appeared to be a precise natural clock (the decay of radioactive elements) and the exactness of mathematics.

Boltwood's work was progressed by a bright young British geologist, Arthur Holmes (1890–1965). In 1911, aged only 21, he embarked on a lifetime's quest "to graduate the geological column with an ever-increasingly accurate time scale". Holmes began dating rocks from several different geological periods, and the ages he calculated ranged from 340 million years (for a

Carboniferous sample) to 1,640 million years (a Precambrian sample). Within a year, he was able to propose the first geological time scale based on radiometric dating. Holmes's initial estimates of Earth's eras have held up remarkably well over time: for example, he placed the beginning of the Cambrian period at around 600 million years ago; today, 545 million years is the time frame that is largely accepted. In the early 1900s, however, Holmes's results appeared to be at odds with other methods in common use, and they were not met with immediate acceptance from all quarters.

Since Holmes's initial work in 1911, many improvements have been made to the process of radiometric dating. Of major importance was the discovery in 1913 that the atoms of a chemical element can exist in two or more different forms, called "isotopes", which are the same chemically, but have different atomic masses. Some isotopes are stable, and it is only the unstable ones that undergo radioactive decay. Of crucial importance for radiometric dating purposes, different unstable isotopes of the same element often have very different half-lives. For example, most uranium consists of the isotope U238, with a half-life of 4,500 million years; however, about 0.7 per cent is made up of the isotope U235, which has a half-life of 713 million years. Lead comes in several stable isotopic forms, some which occur in minerals only through the decay of other, radioactive isotopes, and yet others that were present in the minerals from when they first formed. It became apparent to Holmes and others that, for better accuracy, it would be necessary to measure isotope ratios (eg U238 to its decay product, the lead isotope Pb206), rather than just uranium to lead.

The discovery of isotopes initially complicated the process of radiometric dating, but did in time make it more precise. One initial effect was a reappraisal of Holmes's time scale. As a result of not compensating for then unknown factors, his computed ages were too high. Researchers realized that radiometric methods held promise for reassessing the Earth's age. In 1921, American astronomer Henry Russell (1877–1957) obtained 4,000 million years as a rough approximation of the age of the Earth's crust. This was based on an average of its maximum age calculated from its total uranium and lead content, and a minimum age based on the oldest known (at that time) Precambrian minerals. Over the following years, several more different ages for the Earth's crust were computed and published. These included 3,400 million years (Rutherford, 1929) and 4600 million years (Meyer, 1937).

Meanwhile, older and older rocks were being found in different parts of the world. By the 1940s, it became apparent that, to calculate an accurate age for the Earth, one piece of data was still needed – the ratio of different isotopes of lead in the Earth's crust at the time of its formation. Eventually, in 1953, the American geochemist Claire C. Patterson (1922–95) was able to infer this ratio through measurements on minerals of the Canyon Diablo meteorite, which contains very little uranium.

Meteorites, bits of mineral material orbiting the Sun that have only recently collided with Earth, are thought to have undergone very little change or reworking since their formation. As a result, their radiometric ages should be very close to the age of the solar system – and, by extension, the age of the Earth. Patterson was able to calculate an age for the meteorite of 4,550 million years, give or take about 70 million years. From studying the mixture of lead isotopes in the meteorite, and the same mixture in Earth crustal rocks, Patterson was also able to say that Earth was formed at the same time as the meteorite. Despite various further refinements to radiometric dating since then, his measure of 4,550 million years for the age of the Earth is still regarded as accurate.

During World War II, intense research on the atomic bomb led to major improvements in equipment for identifying and analysing isotopes. It became possible to detect minute quantities of specific isotopes and to measure their abundance with high precision. This in turn has led to highly accurate dating methods.

Modern radiometric dating of rocks is based on a number of different isotope combinations. For dating rocks between one and 100 million years old, an isotope with a shorter half-life is used – potassium 40 to argon 40 (half-life 1,300 million years). The decay of rubidium 87 to strontium 87 (half-life 47,000 million years) is another combination occasionally used.

Radiometric dating could originally only be applied to igneous rock, ie rock formed from the crystallization

of minerals from a molten material (magma) as it cooled. Indeed, until only recently, sedimentary rock was not suitable for radiometric dating. This is because the age of a specific grain in a sedimentary rock, such as a sandstone, is the age at which the mineral formed in its original igneous setting and not when it was locked into the sedimentary rock deposit. To discover the dates of sedimentary rock strata, geologists have traditionally had to find igneous rocks the age of which could be related to layers in the sedimentary strata (through superposition or crosscutting relationships). For instance, dating zircon from sediment illustrates certain problems with radiometric dating. A robust mineral, zircon may

have survived several cycles of erosion and many millions of years before being included in the sediment, making dating using the usual radiometric method inaccurate. A new radiometric method, however, is able to date tiny crystals that grow on zircon grains after deposition. As a result, it is now possible to date some sedimentary rocks where this type of additional mineral growth occurs. However, carbonate minerals which crystallize from carbonate-enriched water in some limestones, especially those deposited within caves, also contain radioisotopes which can be dated. They have proved particularly useful in dating climate changes during the Quaternary ice ages.

Oldest rocks

Early Precambrian (Hadean age) rocks at Isua in Greenland are among the oldest rocks still preserved at the Earth's surface. They were originally formed as volcanic deposits and sediments, mostly deposited in water around 3,800–3,700 million years ago. They have since been metamorphosed but still retain hydrocarbon residues of organic origin, showing that life was present in the waters they were formed in.

The rock record of earth's origins

Meteorite

Meteorites are the only rocks on Earth which have not been formed here but in other parts of the solar system. Many of them are derived from the early formation of the Earth and other planets, consequently, they are invaluable sources of information about the materials and processes of that event.

In searching for Earth's origins, scientists have had to look in the Earth itself, on the Moon and planets, in meteorites (rocks that have come to Earth from space), and even in the gaseous emissions of comets.

Any hypothesis for Earth's formation has to fit within a larger model for how the whole solar system formed. The accepted model, developed in stages over the past 200 years, is called the "solar nebular hypothesis" and was first suggested by Pierre Simon Laplace (1749–1827) in 1796. It proposes that the solar system originated in a huge cloud of gas and dust that gradually condensed into a smaller spinning disc. The centre of the disc collapsed to form the sun, while the planets, asteroids, and comets formed through a particle accretion in the outer part of the disc.

Scientists have long suspected that no Earth rocks survive from its original formation. For over 50 years, however, they have been aware that lying around on the Earth's surface are pieces of the original solar system – namely "meteorites". That meteorites are "rocks from space" was first recognized in the 19th century and, by the middle of the 20th century it was established that most originate from the asteroid belt. They are chunks of asteroids that fragmented, became dislodged from their orbits, and fell to Earth. Analysis of the relative element abundance in meteorites indicated that they formed from the same parent material as the Earth, a fact that strengthened the solar nebular hypothesis. It became clear that meteorites hold clues not just to the origins of the solar system, but also to the origins and age of the Earth.

In the 1950s, scientists began to examine meteorites for a specific purpose. They wanted to find out the composition of "primordial" lead, ie the ratio of different isotopes of lead that existed in the solar system at the time of its formation (isotopes are different forms of the atoms of a particular chemical element). With this data, they hoped to find how much additional lead had been formed from radioactive materials in the Earth's crust since its formation, which would give an accurate age

for the Earth (see "Radiometric dating" p.153). What they were looking for was a meteorite mineral, which contained lead, but no radioactive material. The composition of that lead could be considered as "primordial".

As we have seen, in 1953, an American geochemist, Claire Patterson, found such a mineral in the Canyon Diablo iron meteorite, (responsible for a large crater in Arizona). Using Patterson's data, the Earth's age was roughly computed later that year – by Fritz G Houtermans (b. 1929) – at 4,500 million years ± 300 million years. Patterson went on to date several meteorites radiometrically and found a remarkable level of uniformity: their ages were all clustered around 4,550 million years. Furthermore, his analysis of lead isotopes from meteorite and Earth minerals showed that both shared a common origin. It seemed that, 4,550 million years ago, both Earth and meteorites formed out of the same solar disc material. This provided further support to the solar nebular hypothesis and, at last, a reliable figure for the age of the Earth. Despite various further refinements to radiometric dating, Patterson's figure of 4,550 million years has never been much improved upon.

Apart from meteorites, very little material has been found on Earth that is more than 4000 million years old. The oldest rocks discovered so far are the Acasta gneiss complex found near the Great Slave Lake in Canada (see below), dated at 4,030 million years old, and the Isua supracrustal rocks in Western Greenland, dated at

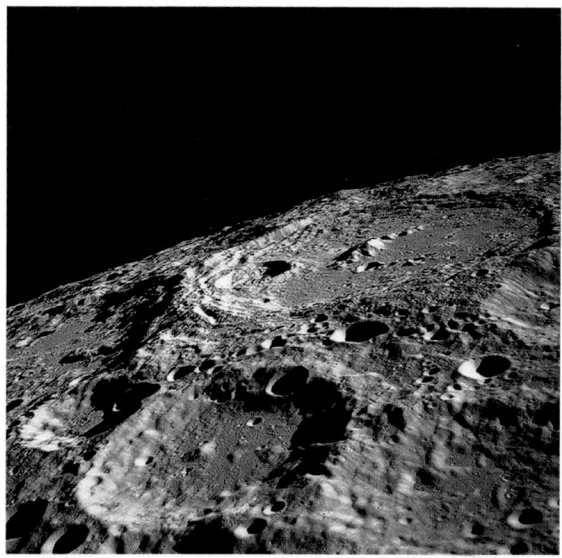

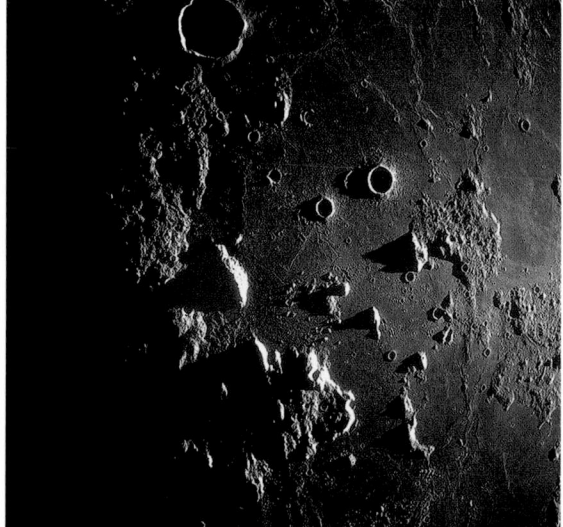

Moon surface

The surface of the Moon is heavily cratered with successive generations of impacts of all sizes and vast lava fields, demonstrating its catastrophic early history after being torn from the Earth around 4,500 million years ago. The Earth also suffered later bombardment from space around 3,900–3,800 million yeas ago but the craters have been reworked by ongoing dynamic geological processes

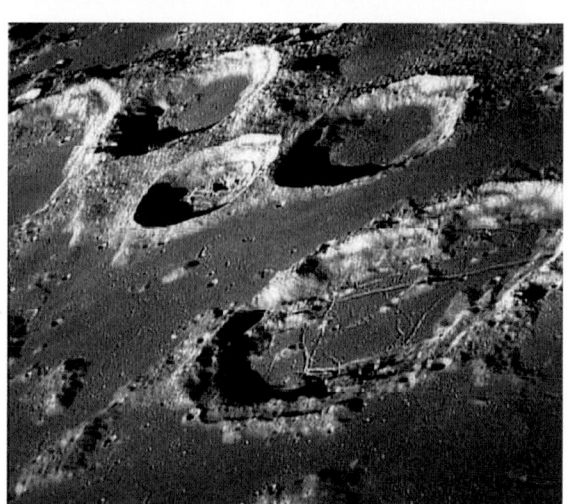

3,700–3,800 million years old. In addition, some zircon mineral grains, dated at up to 4,400 million years old, have been found in metamorphosed sedimentary rocks in western central Australia.

Even assuming (from the zircons) that some type of solidification occurred at Earth's surface some 4,400 million years ago, this still leaves a gap of about 150 million years back to its initial formation. There are good reasons to believe that, for at least part of the time, the planet was completely molten, with evidence for this coming from Earth's overall structure – a core containing the heavy elements iron and nickel, and a mantle and crust, which is dominated by lighter elements such as silicon and oxygen. Such differentiation could only have occurred through

heavier elements "sinking" into the centre at a time when the Earth was liquid.

The early Earth would have melted firstly, because a huge amount of heat was generated when the Earth first formed, through the release of gravitational energy. Secondly, from the beginning, radioactive elements would have been generating heat. Thirdly, the early Earth was bombarded by comets and asteroids, which would have released further large amounts of heat as they hit the surface. The differentiation into core and mantle would itself have released large amounts of heat, so it is reasonable to assume that Earth remained molten for many tens of millions of years after this event.

In the late 1960s and early 1970s, rock samples were retrieved from the Moon but the data obtained from them provoked more questions than answers. The Moon rocks were slightly younger than the oldest Earth rocks, but their chemical (isotope) composition was similar apart from a deficiency of iron. These findings eliminated two theories which, up to then, had been leading contenders for how the Earth–Moon system formed. The two bodies could not have formed together out of the same material – if they had, why would Moon rocks be younger than Earth rocks, and why would the Moon lack iron? And the Moon could not be a stray body that was captured by Earth's gravity. After all, why would it otherwise be so chemically similar to Earth?

In the 1980s, a new theory emerged for the origin of the Moon that seemed to fit the facts. Today, it is accepted as the most likely explanation. According to this theory, about 4,500–4520 million years ago – soon after it first formed and after differentiation into core and mantle – the Earth collided with another object about the size of Mars. As a result, a huge amount of crustal material was flung out and went into orbit about the Earth. Eventually, this material accreted to form the Moon.

Computer simulations have demonstrated that the collision theory is feasible. Such simulations have indicated temperatures of more than 10,000°K immediately after the event. As the ejected material accreted to form the Moon, an early crust formed from rocks called anorthosites (these now formed the lunar highlands), with the large basins known today as *mares* ("oceans") being excavated by impacts, then infilled with molten rock. Evidence from the well-preserved lunar surface indicates that the number of impacts lessened significantly as the solar system stabilized around 4,000 million years ago. The Earth would also have suffered the intensity of this late bombardment, but evidence of this is not preserved because of the dynamic reworking of the Earth's crust by plate tectonics.

If the collision did happen, in some ways it was a fortunate catastrophe because, without it, life on Earth may never have taken a hold. Among other consequences, the Moon eventually stabilized the Earth's rotation – if there were no Moon, the Earth would "wobble" more in its rotation, and we would have much more extreme seasons and weather.

As mentioned above, the oldest materials discovered so far with an identifiable Earth origin are some tiny crystals of a very tough mineral called zircon (zirconium silicate). These were found in 1999 within some sandstone rocks in Australia. They have been radiometrically dated as 4,400 million years old, implying they were formed a mere 150 million years after Earth itself.

The existence of these zircons has some deep implications for Earth's early history. Zircons most commonly crystallize in granite, a type of rock associated with continental crust, and there is every reason to suppose that this is how the Australian zircons also originally formed. Chemical analysis of the zircons using a technique called an ion microprobe has also shown that they contain an isotope of oxygen, oxygen–18, suggesting that they must have been formed in the presence of water.

Overall, the investigation of the zircons indicate that Earth may have developed continents and had some surface water – perhaps even oceans – as long as 4,400 million years ago. This is much earlier than previously supposed; however, they would have been destroyed by the late bombardment and had to re-form.

The zircons formed as crystals within molten granite that was cooling to form solid rock. The zircon-laden granite was eventually thrust upwards to form mountains, which later eroded. The granite vanished, but the zircons ultimately came to rest 3,000 million years ago in sandy Australian stream sediments. These sediments later hardened into rocks that subsequently were altered by heat and pressure.

If there were water on the Earth's surface 4,400 million years ago, the question arises – where did it come from? One possibility is that it was outgassed by volcanoes, but another favoured theory is that water was brought to Earth primarily by comets. Evidence from the lunar surface indicates that the Moon was subjected to intense bombardment from its formation until 3,900 million years ago. During that time, it was ravaged by more than a million major impacts from comets and asteroids. But Earth suffered bombardment, too, which may explain why no rocks survive from the first 500 million years of Earth's history – any crust that formed was soon destroyed by impacts. Also, as roughly half the content of comets is water, cometary impacts could certainly have brought plenty of water to the surface.

The "oceans from comets" theory has suffered some setbacks in recent years from studies of the gas emissions of comets. These indicate that the water in some comets, such as Comet Hale-Bopp, differs significantly from the water on Earth in its content of hydrogen isotopes. The water in other comets, however, such as Comet Linear (which broke up in 2000 as it passed the Sun), is very similar to Earth's. At present, the consensus is that comets may have contributed to Earth's oceans, but were probably not the sole source.

The composition of the Earth's atmosphere during its first few hundred million years of existence is almost entirely the subject of guesswork – there is very little evidence of what it may have contained, least of all in rocks. Some planetary scientists have speculated that the composition of the planet Jupiter today may be representative of the ancient atmospheres of the smaller planets. They base this conjecture on the fact that, because of its size, Jupiter has retained all the light gas

molecules it ever had, and these gases would have been present throughout most of the solar nebula.

On this basis, Earth's original atmosphere may have consisted mostly of hydrogen with some helium, but most of this is likely to have dissipated very quickly into space as a result of heating from the Sun. Earth's gravity would not have been strong enough to retain these light gases. The remaining atmosphere is likely to have contained some slightly heavier molecules formed from the most common elements – likely candidates being carbon dioxide (CO_2), methane (CH_4), ammonia (NH_3), nitrogen (N_2), water (H_2O), and some sulphur gases, but no free oxygen. Some of these gases were probably taken up in large part by dissolving in the oceans once these had formed, and others may have gradually been destroyed by light-mediated reactions. By around 3,500 million years ago, the atmosphere may have consisted mainly of nitrogen and carbon dioxide. Once photosynthesizing organisms appeared in the oceans and then on land, the stage was set for take-up of the carbon dioxide and liberation of free oxygen.

Star formation

The formation of a star such as our Sun begins when part of a nebula begins to coalesce into denser aggregations of gas. As the gas ball shrinks it becomes hotter, with temperatures and pressures at its centre becoming high enough to spark off nuclear reactions that convert hydrogen to helium and generate enough energy to make the star shine.

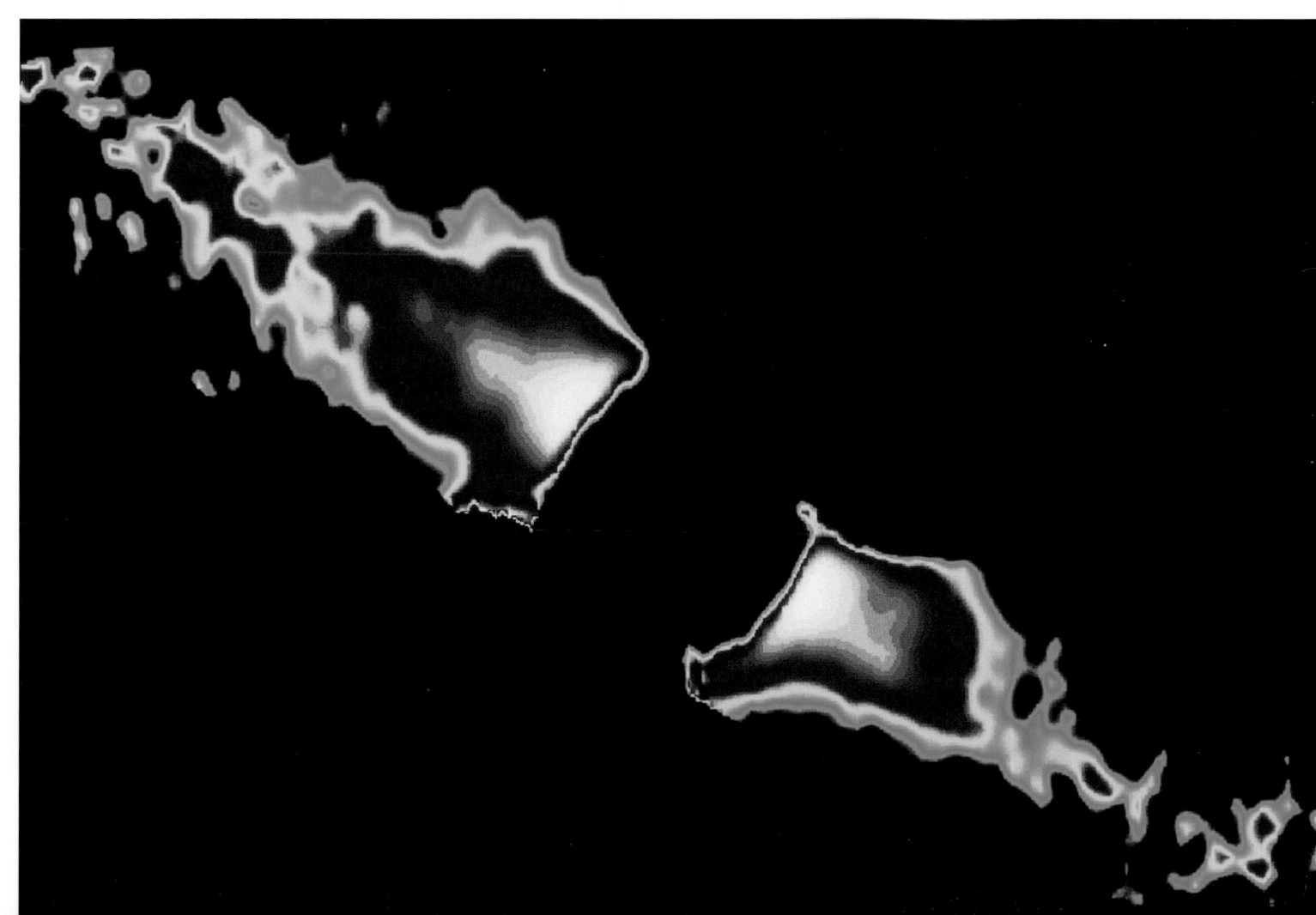

9

The Earth's future

Over recent decades, geological discoveries have shown that the Earth is a much riskier place to inhabit than previously thought. There have been both internally driven processes and external events that occur infrequently but on a much larger scale than anything before imagined. Plate tectonic movements have pushed continents from one hemisphere to another. Major impact events from space have repeatedly wiped out more than 50 per cent of all living organisms. Ice ages and vast outpourings of flood basalts have occurred several times. All these have had major impacts on the life of the past.

The good news is that these events have very low frequencies of occurrence and, despite all these vicissitudes, life has not only survived but has bounced back after the catastrophes. However, life is never quite the same after such an event.

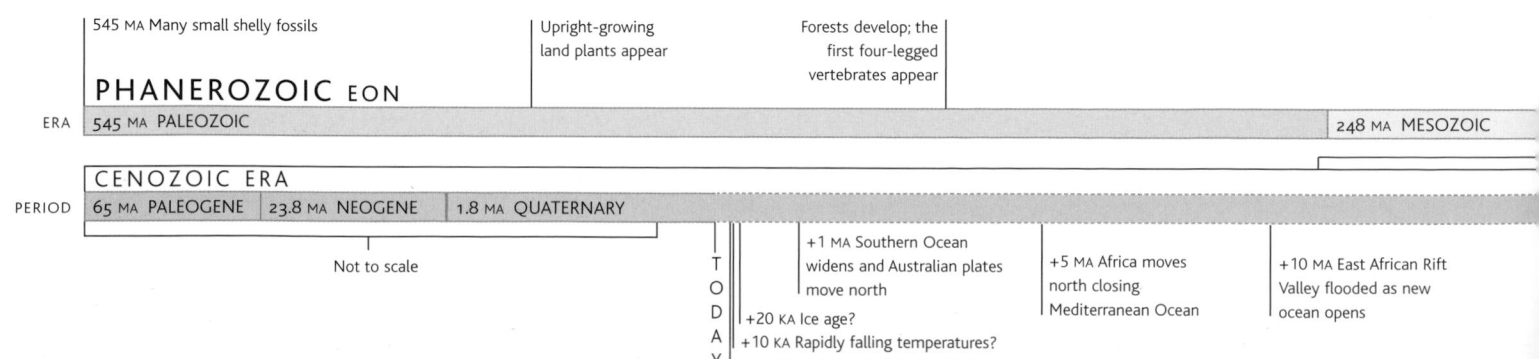

| ERA | 545 MA Many small shelly fossils

PHANEROZOIC EON
545 MA PALEOZOIC | Upright-growing
land plants appear | Forests develop; the
first four-legged
vertebrates appear | 248 MA MESOZOIC |

| PERIOD | CENOZOIC ERA
65 MA PALEOGENE | 23.8 MA NEOGENE | 1.8 MA QUATERNARY |

Not to scale

TODAY

+1 MA Southern Ocean widens and Australian plates move north
+20 KA Ice age?
+10 KA Rapidly falling temperatures?
+3 KA Global warming maximum?

+5 MA Africa moves north closing Mediterranean Ocean

+10 MA East African Rift Valley flooded as new ocean opens

As a result of these catastrophic events, old established groups such as the trilobites, the clubmosses (lycopsids), the dinosaurs, and pterosaurs are eclipsed and new groups, such as the crabs, flowering plants, songbirds, rodents, and humans come into dominance.

The earth sciences have a great deal to say about the Earth's future. From the study of these processes and events of the geological past, their mechanisms and range of magnitude and frequency, it is possible to develop scales of probability for future occurrences in much the same as an actuary calculates life expectancy based on family history, life style, and environment in which the subject lives.

While Earth as a planet has a very long way to go before any signs of "mortality" set in, a large number of highly significant events that will impact upon life will happen, some dramatically sudden, others noticeable over a lifetime, and others still more gradual but that will nevertheless have a very serious affect on life on Earth.

Blue planet

Two-thirds of the Earth's surface is covered with ocean water, while the remaining third is land. Vast areas of white cloud, made of water droplets, further demonstrate the vital role that water plays in the formation of our atmosphere, which protects the land and its inhabitants from harmful solar radiation.

The first dinosaurs and early mammals appear

Birds and flowering plants appear

Primates and songbirds appear

7 MA First hominids appear on Earth

TODAY

65 MA CENOZOIC

+20 MA Widening of Atlantic Ocean

+40 MA Australia crosses the Equator; Antarctica moves north

+50 MA Major impact and extinction event?

Climate change

Geological records show that, over time, climate can change drastically on a global scale. At one extreme, there were times when there were no ice caps, temperate deciduous forests grew at the poles, and subtropical conditions extended north to today's London and New York. At the other extreme, it is possible that much of the Earth, both continents and oceans, was completely frozen over – the so-called –snowball Earth state. This latter extreme, however, was confined to Precambrian times, and is highly unlikely ever to return. Nevertheless, there have been several glacial episodes in Earth's history; we are still living in one. We know that this range of climates can be caused by many factors, from plate tectonics changing the shape of oceans and configurations of the continents, astronomical cycles involving the Earth and Sun, to major episodes of volcanism changing the composition of the atmosphere. This complexity makes the modelling of climate change very difficult, but with today's computers considerable advances have been made.

Geological history also tells us that climate change has had severe impacts on environments and life, whether it be glacial ice-sheet advance bulldozing vegetation and soils from landscapes, or desertification drying up plant life over vast regions. Such drastic changes have occurred frequently in the past and will occur in the future. The critical questions are how frequently and how rapidly changes will take place. To discover this, high-resolution geological records are required. The best are inevitably the most recent – those which cover the past million years or so.

A number of independent sources provide detailed information about climate change during the Quaternary ice ages, especially the last glacial, which ended around 15,000 years ago, and the subsequent deglaciation. We have now been living in an interglacial for around 15,000 years. But how long do interglacials tend to last? Over the past 800,000 years or so, the answer to that is around 11,000–12,000 years – although there is evidence from sediment cores in the North Atlantic indicating that it has been free of ice-rafted debris for some 14,000 years now. So, according to "recent" history, another glacial might well be overdue. There was one interglacial, however, which lasted much longer. It occurred around 400,000 years ago, when the orbital eccentricity of the Earth was minimal, a situation which also obtains now, so there is a reasonable chance that this present interglacial may also last for a few more thousand years.

Apart from uncertainty about when the next glacial is due, we also have to wonder exactly how fast conditions will change when it does happen. Within historic times, we have seen how the Little Ice Age lasted some 350 years from the 15th century into the mid-18th century; before that there was a Medieval Warm phase from 1100–1400. Even so, global average temperatures have been remarkably stable over the past 8,000–9,000 years – around 15.5°C (60°F).

However, if we look at detailed records of the transition from the last glacial (with global average temperatures around 10°C (50°F)) we see some very worrying fluctuations. Between 14,500 and 13,000 years ago, there was a major pulse of meltwater into the oceans, reflecting a significant warming event in the northern hemisphere known as the Bolling/Allerod interstadial. This was followed 13,000 years ago by a

Frozen canals

Historic paintings from northern Europe frequently show frozen rivers and canals from the 15th to the mid-18th centuries, a period known as the "Little Ice Age", when there were famines, major outbreaks of disease and social unrest. The paintings are accurate depictions and sobering reminders of significant climate change within recent times.

PERIOD	1.8MA QUATERNARY		TODAY	
EPOCH	1.8 MA PLEISTOCENE	10.000 MA HOLOCENE	"FUTURE-CENE"	

Climate change through Phanerozoic time

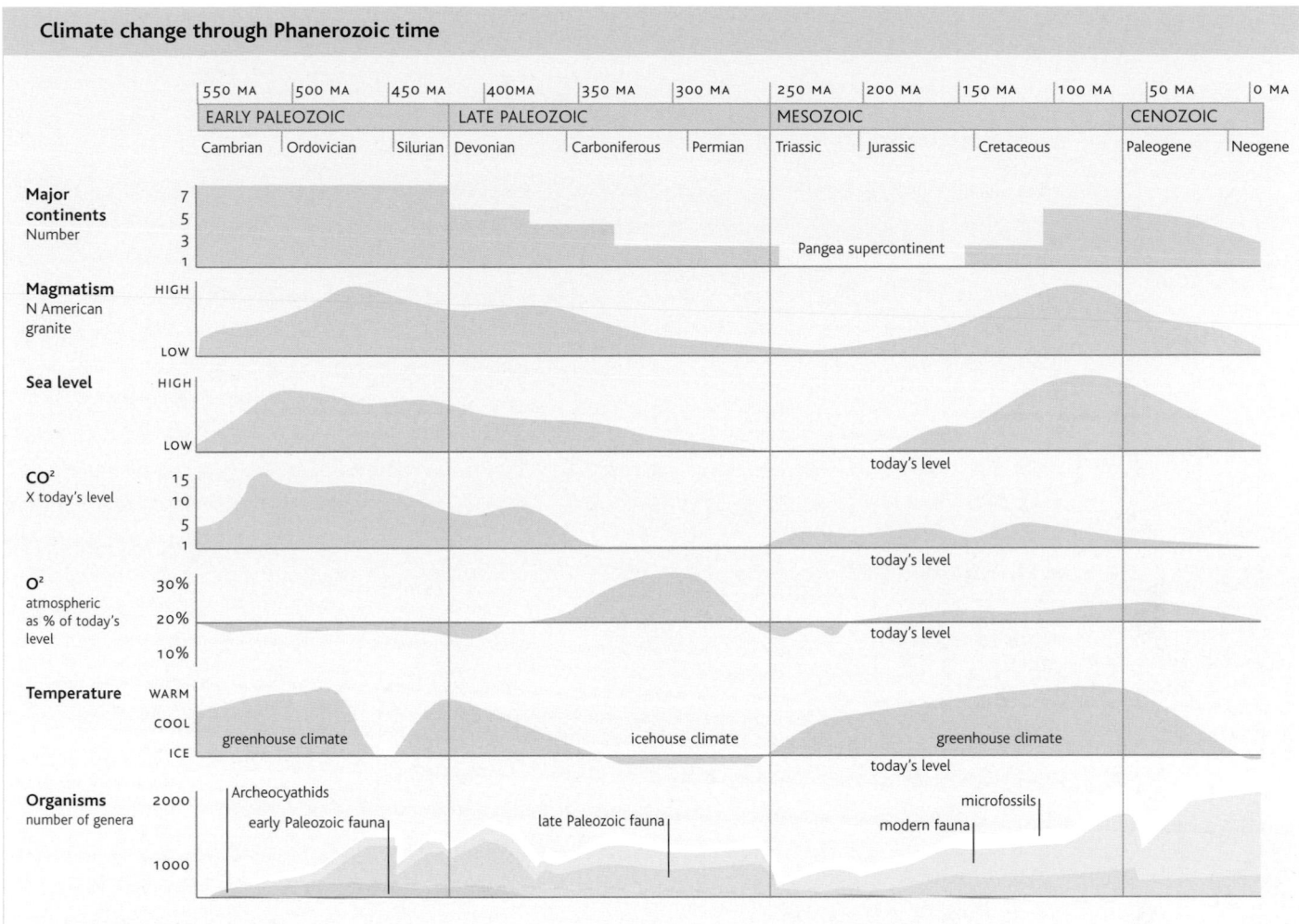

		550 MA	500 MA	450 MA	400MA	350 MA	300 MA	250 MA	200 MA	150 MA	100 MA	50 MA	0 MA
		EARLY PALEOZOIC			LATE PALEOZOIC			MESOZOIC				CENOZOIC	
		Cambrian	Ordovician	Silurian	Devonian	Carboniferous	Permian	Triassic	Jurassic	Cretaceous		Paleogene	Neogene

Major continents Number — 7, 5, 3, 1 — Pangea supercontinent

Magmatism N American granite — HIGH, LOW

Sea level — HIGH, LOW

CO_2 X today's level — 15, 10, 5, 1 — today's level

O_2 atmospheric as % of today's level — 30%, 20%, 10% — today's level, today's level

Temperature — WARM, COOL, ICE — greenhouse climate, icehouse climate, greenhouse climate, today's level

Organisms number of genera — 2000, 1000 — Archeocyathids, early Paleozoic fauna, late Paleozoic fauna, modern fauna, microfossils

sudden cold snap in the northern hemisphere known as the Younger Dryas cooling event. From the data, it looks as if within a hundred or less years the global average temperature dropped several degrees only to be followed by an equally rapid rise into the present interglacial. Not surprisingly, this is one of the most studied of past climate events because it is so dramatic. Looking back deeper into the ice age, it also looks as if many of the earlier climate changes might also have been just as rapid. If this is true, then we – or our descendants – are going to be in big trouble.

Terrestrial plants, the basis of our food chain, do not respond well to decadal scale changes in climate and much of our essential food crop agriculture would be severely damaged. The only good result will be the reversal of the present rise in global sea levels as more ocean water is locked up on land as snow and ice.

Consequently, the frequency of disastrous flooding of low-lying, densely populated, and intensely farmed tropical regions will also be slowed.

We know that present global warming is enhanced by human (anthropogenic) activity and the pumping of greenhouse gasses into the atmosphere since the beginning of the Industrial Revolution. Atmospheric carbon dioxide levels have risen from 280 parts per million (ppm) to 365ppm in less than two centuries. So will this not put off any return to glacial conditions? The answer is probably not, because global warming is simply adding more uncertainty to the situation. As warming proceeds, the polar ice sheets and glaciers are melting, releasing more lower-density freshwaters to the oceans. As ocean water warms, its density decreases, and this may well disturb its global circulation in an unpredictable way, with severe consequences for the climate.

Shifting continents

Oceans have been opened and closed by the processes of plate tectonics ever since the differentiation of the Earth's core mantle and crust some 4,000 million years ago. As long as the internal mechanisms that drive plate tectonics continue to work, the Earth's continents will continue to be shuffled around on the surface of the planet in ever changing configurations. Rising plumes of heat and slow plastic flow of mantle rock materials generate ocean-floor spreading ridges and hotspots. Crustal plates are heated, expand upwards, break apart, and slide away. New ocean floor is created and just as much is destroyed by subduction because the Earth cannot expand.

Predictable future movements of the continents will depend on continuing patterns and rates of plate motion. At present, the Atlantic and Southern oceans are actively spreading at rates of around 10mm (less than ½in) a year. This may not seem much, but it soon adds up over geological time – to 10km (6 miles) per million years. Consequently, the "pond" between the Americas and Europe and Africa will widen. The net effect will be that the Americas will move further and further westwards into the Pacific. If the process continues in this way, the motion of the Americas will close the Pacific, subducting the ocean floor and generating huge amounts of volcanism along the western, leading edge of South America.

As Africa is surrounded to the west, south, and east by spreading ridges, its net movement will continue to be in a northward direction. The Mediterranean remnant of the Tethys Ocean will be closed, and the Alpine mountain belt will be reinforced. The classical world of Greece and Italy will be pushed and crumpled into newly formed southern alps, rather like a car going through a crushing machine. Northeast Africa will break away from the rest of the African continent and head off northeastward. Madagascar will go with it, and together they will move towards India and southeast Asia as the Great East African Rift Valley opens up into a new ocean. India will continue to move slowly north, maintaining the elevation of the Himalayas and Tibetan plateau behind it.

Australia will further distance itself from Antarctica and move rapidly north across the equator, sweeping the Indonesian islands ahead of it and perhaps leaving New Zealand behind. In 50 million years, most of the Australian continent, apart from the southeast and Tasmania, will have crossed the equator and, by 100 million years from now, it will have crashed into the Japanese islands and rammed them into China, forming a huge new mountain range. Thus, 100 million years hence, there will be an even greater concentration of continents in the northern hemisphere, forming a vast new supercontinent, just as Gondwana and Pangea were formed in the Paleozoic era. The big question will be what happens to Antarctica: will it be left at the south pole or will it join the other continents?

This supercontinent will not last very long because the insulating effect of such an enormous continental "blanket" on the underlying mantle will concentrate heat flow, leading to doming volcanism, with the effusion of large volumes of flood basalts and rupture which will rift the supercontinent once again. And so the process will continue, we know not where.

Of all this, the most drastic short-term (in geological terms) effects will be in the Mediterranean, where earthquakes, volcanism, and mountain building will intensify. The same processes will also impinge upon life in northeast Africa as it rifts away, in the Japanese and Indonesian islands, and along the western margin of South and Central America as subduction continues. Anyone living in Australia will certainly feel the effects of its rapid northward motion. The interesting thing will be how the climate changes and how extensive the tropical rainforests will become as the continent crosses the equator and then returns to aridity as it passes into the Tropic of Cancer. Western North America's future will not be smooth either, as California continues its northward motion along the San Andreas Fault and slides up the west coast, past Canada, towards Alaska.

At another scale of magnitude, the future prospects of the Earth are intimately tied up with that of the Sun, which is scheduled to die in some 5,000 million years. By then, the Sun's central hydrogen will have been used up. It will swell into a red giant (a very large star of high luminosity and low surface temperature), large and bright enough to engulf the inner planets in the solar system. Although the Earth will not be "absorbed", all life on the Earth's surface and in the oceans will be thoroughly cooked, then vaporized. The Sun's red giant phase will last for some 500 million years while the remaining

hydrogen shell burns away around a helium core. Things will then accelerate and become even more dramatic, with fusion of helium in the core which will blast off some of the outer layers (about 25 per cent of the mass). The residue will be a white dwarf (a small, very dense star), no larger than the Earth, which will continue to glow no brighter than a full Moon with a bluish hue, a "stellar cinder" shining on what remains of the solar system. Altogether, the Sun has more time ahead than behind it, and its future history hardly need worry us just now because we have more pressing concerns.

CONTINENTAL CONFIGURATIONS OF THE FUTURE

50 million years from now

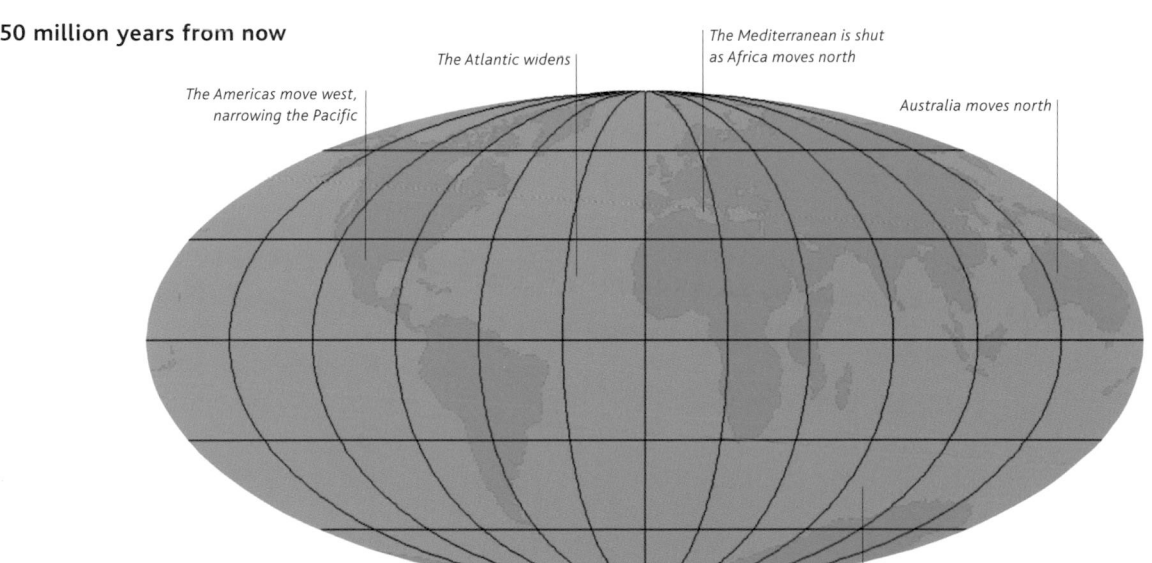

The Americas move west, narrowing the Pacific

The Atlantic widens

The Mediterranean is shut as Africa moves north

Australia moves north

The Southern Ocean widens

Emerging mountains
Africa is moving north and will eventually squeeze the Mediterranean sea down and crash into southern Europe. In 50 million years time the result of the crash will be a vast new mountain range which will make the Alps look small. Italy, Greece, and the remains of the classical world will be crunched up within the mountains, which will be of Himalayan dimensions.

100 million years from now

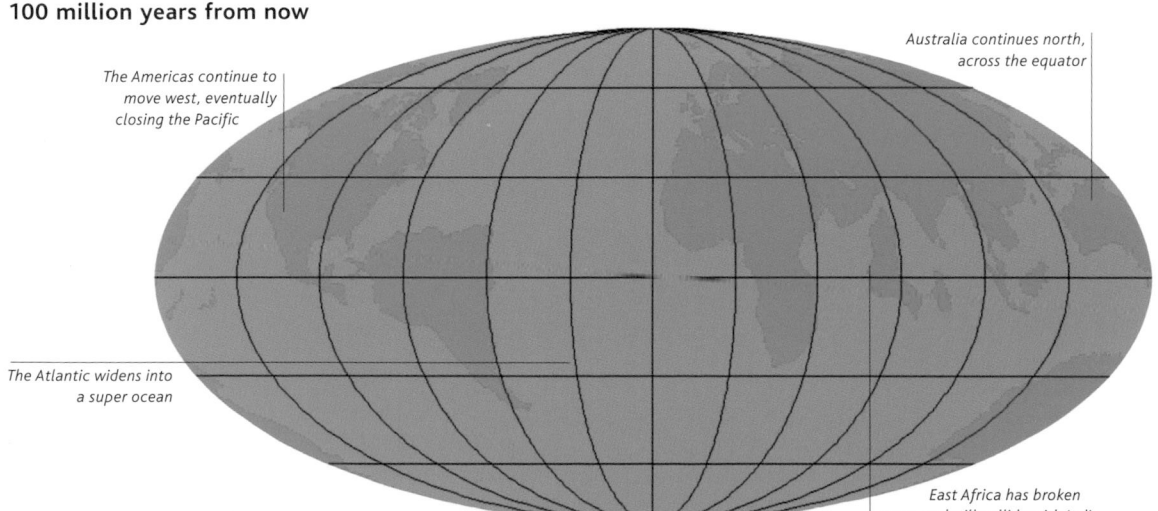

The Americas continue to move west, eventually closing the Pacific

Australia continues north, across the equator

The Atlantic widens into a super ocean

East Africa has broken away, and will collide with India

Australia moves north
100 million years hence, on the other side of the globe, Australia is steadily moving north. It will push past the islands of southeast Asia sweeping some of them along as it heads across the equator towards Japan. The Atlantic Ocean will continue to widen.

A material world

There is a general appreciation that we still need to obtain geological materials such as oil, coal, diamonds, iron ore, lead, and salt from the rocks of the Earth or recycle them whenever possible. We depend upon some 70 minerals, mostly metals, for the running of our modern world. In addition, we use vast tonnages of rock for the construction of roads and buildings. Many of these geological resources have finite supplies, and we need to start thinking about how best to manage them.

Market forces tend to rule supply and demand, but some materials, such as fossil fuels, radioactive materials, and metals also have strategic value. Continuing high demand and diminishing supplies of certain critical earth materials such as hydrocarbons and potable water may well fuel border disputes and even wars in the future. Unfortunately, the geographical distribution of essential raw materials is very unequal because of the nature of geological processes. We are much more dependent upon these materials than most people realize.

Some geological materials are rare and exotic, such as arsenic, with a small annual global production of around 31,000 tonnes, mostly used in herbicides and rodenticides, half of which is mined in China. Then there are more everyday materials such as bauxite ore, which is needed for the production of aluminium. Around 140 million tonnes of bauxite are quarried each year, with more than a third coming from Australia. In terms of sheer bulk, coal is still top of the list, with more than 4,000 million tonnes mined each year and Russia being the biggest single producer (248 million tonnes). Crude petroleum is fast catching up with this, however, with 3,500 million tonnes produced each year, and Saudi Arabia is the biggest single producer, (441 million tonnes).

A few of these raw materials, such as coal, are quite widely distributed; many others are much more localized. Nevertheless, they all require geological processes to concentrate them sufficiently to make their exploitation economic. Lead, with its low melting point

Aral Sea
Since 1960, the Aral Sea in Central Asia has shrunk by 40 per cent, lost most of its once abundant fish stocks and become extremely saline and polluted. Rivers which fed it have been diverted for irrigation and industry, generating an ecological disaster in an area once famed for its wildlife.

THE RECENT PAST

8 KA		1.8 KA	PRESENT
First metal use (lead)		Roman lead smelting	Hydrocarbons will run out

(328°C (622°F)), was among the first metals to be exploited by humans around 8,000 years ago. Roman lead smelting 1,800 years ago and its atmospheric pollution can be picked up in ice cores from Greenland. Today, lead production is steadily declining – to around 3 million tonnes per year because its toxicity has reduced its usefulness. It was once an additive in paint and petroleum. The average abundance of lead in continental crust rocks is 0.0008 per cent, and its value is around US$540 per tonne, so it has to be concentrated (by a factor of 6250 to 5 per cent) by some geological process to make its mining worthwhile. By comparison, gold is much rarer in continental crust rocks (around 0.0000003 per cent). However, it is so highly valued (around US$11 million per tonne) that its extraction is economically viable even when its concentration is so low that it is not visible (around 0.0004 per cent).

Natural concentration varies because of processes ranging from igneous activity that can fractionate rare elements and minerals, such as primary hydrothermal mineralization, to secondary processes, such as weathering and selective transport. For instance, prolonged tropical weathering reduces most igneous rocks to just a few residual minerals such as iron and aluminium oxides forming laterite and bauxite, and the latter often contains as much as 30 per cent aluminium. Some precious metals, such as gold and platinum, do not oxidize or otherwise weather, although they may be physically abraded. Likewise, some valuable gem minerals, such as diamond, ruby, and sapphire (corundum) are very hard and resist weathering and erosion. They all may be concentrated as pebbles and gravel in stream alluvium, and occasionally offshore, in layers known as "placer deposits". Many diamonds are now dredged from placer deposits off the Namibian shore in southwest Africa.

Most mining processes produce an enormous amount of waste rock material, and many require chemical processing which in turn often produces toxic liquid and gaseous waste. For instance, a tonne of ore from a typical copper deposit produces 30kg (66lb) of ore concentrate, from which 8kg (18lb) of refined metal is smelted, leaving 970kg (2,140lb) of waste rock material. Consequently, the processes of extraction and

refining do not sit easily with dense human populations or with aesthetically valued natural landscapes such as those set aside for national parks. Affluent and often crowded Western countries now find the presence of these mineral industries unacceptable. As a result, the industries now tend to be located in less densely populated or poorer areas of the world – although many in the West are now waking up to the ethical problems posed by the exportation of their "dirty" industries. The result is likely to be higher prices in future.

Prices of many raw materials will also rise because of decreasing supplies. For example, oil production is forecast to peak at around 70,000kb/d (thousand barrels per day), or 9.5 million tonnes a day, sometime around now. The rate of discovery of giant oil fields has declined drastically since 1962, with fewer coming on stream to boost future supplies. One result will be a serious price hike as demand outstrips supply, and then, around 2010, it will become significantly worse as supply begins to fall. By 2020, output is predicted to fall to the 1970 level and, by 2050, to the 1960 level; however, demand for oil has not decreased – we have a real problem ahead. While development of oil fields in vulnerable "virgin" landscapes such as Antarctica and northern Alaska is banned through concern for the environment and wildlife, there is going to be a scramble for what remains. Predictably, the Middle East, with its remaining oil riches, is going to be the site of future conflict.

Llannon coal mine
The exploitation of most earth materials of economic value are both bound by market forces and finite supplies. Here in South Wales a coal mine in Carboniferous strata, once part of an active and extensive coal field, has been abandoned because of the cost of extraction and diminishing output.

A dangerous world

Kobe earthquake, Japan 1995

Recent geological investigation has shown why some regions of the Earth are more subject to earthquakes and volcanoes, but accurate prediction is still some way off.

The abundance of living organisms on Earth, including 6,000 million or so humans, would seem to suggest that, despite life's hazards, our planet is a relatively benign place to live. Despite the hazards of reproduction, being born as a relatively defenceless juvenile, the difficulty of obtaining enough food to survive, and avoiding or surviving disease and pestilence, organisms from algae to zebras thrive on planet Earth. Certainly, compared with the other planets in our solar system, Earth is hospitable. The main reason why this is so, however, is a combination of astronomical accidents.

Earth is neither too big nor too small, nor is it too near or too far from the Sun. Its orbit, axis of rotation, and rate of rotation all provide a narrow window that allows the development of a protective atmosphere, hydrosphere, and climate. This climate is generally not too extreme and is thus conducive to life ranging from the microscopic to the blue whale, a 30m (100ft) giant which is the biggest animal ever to have lived.

Furthermore, vast amounts of heat energy are required to keep our cosy little planet going. This energy is transmitted from deep within the Earth and transformed through some rather violent natural processes such as volcanisms. The distribution in time and space of the heat flow is uneven, with the result that some parts of the Earth's surface are a lot more geologically active than others. Fortunately, much of the action takes place on the ocean floor, where new ocean-floor rocks are generated, and in subduction zones, where old ocean floor is returned back to the depths. Still, a lot of the action does impinge upon life on land.

Earthquakes and volcanic eruptions are familiar enough phenomena, even though they are still none too predictable and the pressures of economics and population growth result in very large numbers of people living in regions that are known to be hazardous, such as Mexico City, Naples, and San Francisco. Anyone looking at an earthquake distribution map might perhaps think twice about going on holiday to the Greek islands, parts of Japan, Turkey, or California. And yet we still do so, largely because the frequency of events that are truly catastrophic is low within a decadal scale. So we take the risk, just as we risk travelling by car or taking part in hazardous sports. In the United States, the average risk of death to an individual over a 50-year period is 1 in 100 from an auto accident, 1 in 5,000 from electrocution, 1

THE FUTURE

PRESENT + 10MA	+50MA	+100 MA
Major climate change	Major release eruption	Major impact

in 20,000 from a plane crash, 1 in 25,000 from a hurricane, 1 in 130,000 from lightning, and 1 in 200,000 from an earthquake. If you live in particular earthquake- or hurricane-prone regions, however, then the risks are higher – the figures are averaged out for the whole country. Global risks of death from a volcanic eruption are 1 in 30,000, but this figure is not very meaningful because it entirely depends on where you live, whereas many people now run the risk of suffering an auto accident.

Recent decades of geological investigation have uncovered a number of unexpected hazards for us to worry about. Fortunately, many of these have very low frequency rates, so low that they have not occurred within historic times. But they have happened and will happen again – it is just a matter of when. The fossil record tells us that, on a number of occasions in the past (at intervals of between 100 and 50 million years), life has been drastically cut back (as much as 70 per cent of all species) by extinction events. As far as we are able to tell, there are a number of different reasons for this. They range from a major impact event, such as that which occurred 65 million years ago at the end of Cretaceous times, to combinations of climatic and environmental change such as seem to have occurred at the end of Permian times 248 million years ago.

All the planets in the solar system suffer such impact events from extraterrestrial bodies over time. The most impressive of recent times was the one which hit Jupiter in 1994. Caroline and Eugene Shoemaker and their colleague David Levy spotted a string of 21 fragments approaching the planet in March 1993, at the Mount Palomar Observatory near San Diego, California. Known as Comet Shoemaker-Levy 9 (SL9), the fragments were estimated to be around a kilometre in diameter, and their estimated date of arrival on Jupiter was mid-July 1994. The world was waiting and was not disappointed: the impacts were visible from Earth even with small astronomical telescopes. Just one of the fragments (G), travelling at 60km (37 miles) per second, created an explosion as big as the K/T event. It was equivalent to some 100 million megatons of TNT and produced incredible thermal radiation in the stratosphere that would have fried anything living on the surface. The same would have applied to a similar event on Earth.

The global risk of death from an asteroid impact is 1 in 20,000, which in some senses is a more useful statistic than those for earthquakes for example, as the next asteroid hit could be anywhere on Earth (unlike earthquakes). Still there is no need to panic unduly. The statistic is based on the present total human population and the fact that although 500m- (1,640ft-) wide impacts occur every 10,000 years or so, 1km (³/₅ mile) asteroid hits are much rarer (every 100,000 years or so). However, the latter are likely to kill many hundreds of millions of people when they do impact.

Regrowth
Following Mount St Helen's most recent eruption in 1991 eruption soils have begun to reform on the mineral-rich volcanic debris and plants have begun to re-establish themselves. This volcano in the northwest of the United States is just one of many dangerous volcanoes which lie in a ring around the Pacific Ocean.

conclusion

While the study of the Earth – geology – had its heyday in the 19th century, a new synthesis has refocussed concerns about the future of the planet and our place on it. Fragile natural environments and finite resources have resurrected a sense of stewardship for the Earth, at least in some hearts and minds.

The last few decades have seen a revolution in our knowledge of the geological workings of our planet Earth and the other planets in the solar system. Less than 200 years ago, mankind was still very much at the centre of our world view, although the Earth was no longer seen as the centre of the universe. The Darwin/Wallace theory of evolution, the "new German criticism" of the Bible texts, Marxist theory of revolution, and Freudian theory of the mind and consciousness all helped us to question our view of ourselves and our world. The emergence of post Mendelian genetics, leading to the mapping of the human genome, has continued our reassessment of what we are and how we came to be what we are.

Recent developments in geology have further upset our already shaky certainty and confidence in ourselves as god-ordained "masters of all we survey". The ancient, pre-biblical idea of apocalyptic catastrophism has re-emerged after being submerged by the weight of Lyellian and Darwinian gradualism. While Darwinian evolution displaced us from our self-imposed position above all else except the angels and heaven, it did require that all changes be slow and gradual over many millions of years.

The old catastrophism of a world created by *sturm und drang*, with violent earthquakes, volcanic eruptions, floods, and the cataclysmic generation of mountains was done away with. These events still occurred but were understood to have had little long-term impact on the progressive evolution of life on Earth. However, reassessment of geological evidence for the pattern of progress showed that there were major changes and extinctions in past life. Better understanding of the early evolution of the solar system reinvoked notions of major catastrophic events. The realization that Earth's surface would look like the Moon but for geological reworking of the surface rocks led to the understanding that large rock and ice bodies from space must hit the Earth just as often as the Moon. Better resolution of the fossil record showed that there were major extinction events in the past which seriously damaged life.

These neo-catastrophic ideas came together with the claim that the end of Cretaceous extinction was caused by a major impact event. "Cold war" nuclear-winter scenarios were reworked to explain how such an impact might have caused a global extinction. What role have such infrequent and unpredictable events played upon the course of evolution? Has life been so set back at times that the evolutionary outcome was a matter of chance? If this is true then perhaps another big nail has been banged into the coffin of human self esteem. But is it true?

One certainty is that the future is in many ways uncertain. Deeper geological investigation should help our preparedness for a bumpy ride. Both the planet and the life forms on it have proved remarkably resilient. Despite large rocks repeatedly impacting from space, major glaciations, large-scale volcanism, and constant climate change, life has bounced back and will probably do so for the forseeable future.

glossary

albedo effect The ratio of the amount of solar radiation reflected from a surface to the amount received by it.

altricial Born in an undeveloped state and needing parental care and feeding.

ammonite see *ammonoid*

ammonoids An extinct group of cephalopods (with some 2,000 genera). Early Devonian–end Cretaceous.

amniotes A group of higher vertebrates whose embryos are enclosed in a foetal membrane called an amnion, including reptiles, birds, and mammals.

angiosperms A major group of plants (the flowering plants), with some 220,000 living species, having true flowers with seeds enclosed in an ovary. Late Jurassic or early Cretaceous–extant.

anoxic Lacking in oxygen.

anthropogenic Chiefly environmental pollution or pollutants caused by humans.

arthropods Perhaps the biggest and most diverse group of invertebrates with several extinct fossil subgroups, characterized by segmented bodies, jointed appendages, and an exoskeleton or carapace of chitin, reinforced in some groups by calcite. Early Cambrian–extant.

bacteria General term for prokaryotic, unicellular micro-organisms that lack a nucleus. Probably among the earliest lifeforms in early Precambrian times.

bangiomorphs A Precambrian fossil group thought to be related to living marine red algae.

basalt Dark coloured, fine grained, and silica-poor volcanic igneous rock.

belemnites An extinct group (about 250 genera) of coleoid cephalopods related to the modern squid and octopus. Late Devonian–Holocene.

benthos Organisms that live on or in the sea floor.

biota The animal and plant life of a particular region, habitat or geological period.

biozone A unit of time characterized by a specific assemblage of fossil organisms.

blasto- Prefix denoting a rock texture that has been modified during metamorphism but is still recognizable.

brachiopods A group of marine shelled invertebrates, sessile animals that secrete an external bivalved shell of unequal size and are usually attached to the sea floor by a stalk. Early Cambrian–extant.

bryozoans A group of marine organisms with a calcareous or organic skeleton built into encrusting, branching, or fanlike structures, forming a colony a few centimetres across. Early Ordovician- extant.

carapace The hard shell of a tortoise or toughened exoskeleton of some arthropods such as trilobites and crustaceans.

cephalopods A major group of highly organized marine molluscs of which nautilus, squids, argonauts, and octopuses are the only living forms, and the ammonoids and coleoids are fossil groups. Often with bodies contained within a conical shell and characterized by a head with eight, 10 or more tentacles around the mouth. Late Cambrian–extant.

cetaceans A group of marine mammals characterized by a streamlined, hairless body, no hind limbs, a horizontal tail fin and a blowhole on the top of their head for breathing. Living examples are whales, dolphins, and porpoises. Eocene–extant.

chemosynthesis The "manufacture" or synthesis of organic compounds by bacteria or other living organisms using energy derived from inorganic chemical reactions, typically in the absence of sunlight.

chert (synonymous with flint) A dense, extremely hard siliceous sedimentary rock, consisting mainly of microscopic interlocking quartz crystals.

chondrites Primitive stony meteorites, having crystallised some 4,700 Ma ago and making up over 80 per cent of meteorite falls.

chordata Those animals that at some time in their development have a notochord and gill slits. Mid Cambrian–extant.

cirque, corrie or **cwm** A mountainside rock basin with steep walls excavated by the erosive activity of a mountain glacier.

continental margin The submarine fringe extending between the continental shoreline and abyssal ocean floor. It includes the Continental Shelf, Borderland, Slope and Rise.

continental shelf A gently sloping, shallow water platform extending from the coast to a point (the Shelf Break) where there begins a comparatively sharp descent (Continental Slope at around 20 degrees) to the ocean floor

crustaceans Those arthropods characterized chiefly by two pairs of antenna-like appendages in front of the mouth and three pairs behind it. Present groups include shrimp, crabs, and lobsters. Most forms are marine. Mid Cambrian–extant.

cycads A group of primitive palm-like plants of tropical and subtropical regions, bearing large male or female cones. Early Triassic – extant, but in decline.

cynodonts A group of fossil carnivorous reptiles with well-developed, specialized teeth and ancestral to the mammals. Late Permian–Jurassic.

DNA Deoxyribonucleic acid, a self-replicating material which is present in nearly all living organisms as the main constituent of chromosomes. It is the carrier of genetic information.

dropstones Large rock fragments dropped by floating ice into marine sediments or varved lake deposits.

drumlin An elongated mound of unconsolidated glacial material, commonly occurring in swarms. Up to 60m (200ft) high and several hundred metres long, they are formed under ice sheets or very broad valley glaciers.

echinoderms A group of marine invertebrates having an endoskeleton composed of numerous porous calcite plates, many of which have a pentaradial symmetry such as echinoids and starfish. Late Cambrian – extant.

Ediacarans An extinct group of soft-bodied organisms whose exact biological affinities are not clear. Late Precambrian.

esker A long, narrow, sinuous ridge formed of stratified glacial meltwater deposits, usually with large amounts of sand and gravel.

eukaryote An organism consisting of a cell or cells in which the genetic material is DNA in the form of chromosomes contained within a nuclear membrane. They include all living organisms except eubacteria and archaea.

exotic terranes Tectonic plate fragments.

extremophile A micro-organism, especially an archaean, that lives in conditions of extreme temperature, acidity, alkalinity, or chemical concentration.

foraminiferans or **forams** A very abundant group of amoeba-like marine unicells, many of which secrete calcareous shells. Early Cambrian–extant microfossil found in marine sediment.

gastropods A large and ancient group of molluscs, most of which are shelled and aquatic, both marine and freshwater but includes terrestrial forms. Modern varieties include limpets, snails, and slugs. Early Cambrian–extant.

glacial erratics Ice-transported boulders carried far from their original sources and dumped wherever the ice finally melted.

graptolites a group of marine colonial organisms, related to the living pterobranch hemichordates. Most are extinct and lived within branched organic skeletal tubes. Late Cambrian–Carboniferous with just a few surviving forms.

gymnosperms A large group of plants including ferns and conifers whose seeds are commonly held in cones or other modified shoots and whose ovules are not totally encased by tissue. Late Devonian–extant.

igneous rock Rock formed by the crystallization of minerals from a cooling molten material (magma).

invertebrates All those animals that lack a backbone.

isostasy The condition of equilibrium whereby the Earth's crust is buoyantly supported by the semi-plastic solid rock material of the mantle.

isotope A particular atom of an element that has the same number of electrons and protons as the other atoms of the element but a different number of neutrons.

kettle hole A depression in glacial outwash deposits, formed by the melting of a separated mass of buried glacial ice and the collapse of the overlying sediment.

(K/T) boundary Cretaceous/Tertiary boundary.

lissamphibians The large group to which all living amphibians belong (with over 4,000 species) and separate from extinct amphibian groups. Early Triassic –extant.

magma High-temperature molten igneous rock.

magnetite A strongly and naturally magnetic iron oxide mineral.

mantle A thick solid layer of rock within the Earth extending from below the crust to the core.

marsupials A group of primitive mammals that give birth to small immature young carried and suckled in a pouch on the mother's belly. Early Cretaceous–extant.

metamorphism The processes that produce structural and mineralogical changes in any type of rock in response to physical and chemical conditions differing from those under which the rocks originally formed.

monotremes A group of primitive mammals that lay large yolky eggs. Mid Cretaceous–extant.

moraine Rock debris that has been carried and deposited by a glacier.

notochord A cartilaginous dorsal stiffening rod that supports the body of all embryonic and some adult chordate animals. Precursor to the vertebral column.

obsidian Fine grained and rapidly cooled volcanic glass, usually black but sometimes brown or red.

orogenic belt A linear or arcuate mountainous zone in the Earth's crust, characterized by deformed and metamorphosed rocks, frequently associated with large deep plutonic intrusions and surface volcanoes.

peneplain A hypothetical surface, normally close to sea level, to which landscapes are reduced through prolonged mass wasting, stream erosion and sheet wash.

permafrost Soil or subsoil that is permanently frozen.

petrification or **permineralization** The preservation of hard parts of many organisms (both plant and animal) by mineral-bearing solutions after burial in sediment.

phylum (s), **phyla** (pl) Principal taxonomic category that ranks above class and below kingdom.

placentals A group of mammals that develop a special tissue, the placenta, by which embryos exchange nutrients and waste products with their mothers until they are born. Mid Cretaceous–extant.

placer deposit Superficial sedimentary deposit laid down by water, containing economic quantities of valuable minerals.

planetismal A body of rock and/or ice formed in the primordial solar nebula, from which larger planetary bodies are thought to have formed by coalescence.

plate tectonics A synthesis of geological and geophysical observations in which the plates of the Earth's outer rigid lithosphere move relative to each other, diverge to form new oceans, converge to form volcanic island arcs and mountain ranges and slip past one another as transform faults.

pluton Any major intrusive body of igneous rock formed deep within the crust by partial melting and then slow cooling of magma.

precocial Those young hatched or born in an advanced state and able to feed themselves almost immediately.

prokaryotes Those primitive micro-organisms in which the genetic material is not contained within a discrete membrane-bounded nucleus, but scattered throughout the cell.

reptiliomorphs A group of extinct and primitive reptile-like tetrapods. Mid Carboniferous–early Triassic.

RNA Ribonucleic acid, a nucleic acid found in all living cells.

roche moutonnée A glacially shaped mound on a glaciated bedrock surface.

rostroconchs A group of primitive extinct bivalved molluscs. Earliest Cambrian–end Permian.

rudists A group of extinct bivalve molluscs, many of which had large, conical, coral-shaped shells and formed reef-like structures. Cretaceous.

scaphopods A small group of marine univalved molluscs (about 1,000 species) with hollow curved tusk-like shell, open at both ends. Middle Ordovician – extant.

sedimentary rock A rock formed by the burial and consolidation of sediment settled out of water, ice or air and accumulated on the Earth's surface under the influence of gravity, either on dry land or under water.

speciation The evolution of different species.

stomata The leaf pores plants use to regulate their "breathing" (gas exchange).

stratum (s), **strata** (pl) A layer or series of layers of sedimentary rock generally separated by bedding planes which approximate to the horizontal when the sediment was originally deposited but may since have been inclined or folded by earth movements.

stromatolites Laminated and mound-shaped calcareous sedimentary structures produced by alternating microbial films and thin layers of sediment.

subduction The movement of one crustal plate, generally the denser, under another, with the descending plate sinking into the mantle and eventually being consumed at depth.

taxon (s), **taxa** (pl) One of a hierarchical group of organisms ranging from species to kingdoms.

Tertiary period Original term for life's third age, now divided into two periods – Neogene and Paleogene.

till Generally non stratified sedimentary rock debris deposited directly by glacial ice.

tillite A sedimentary rock formed by the compaction and cementation of till.

trilobites An extinct group of marine arthropods characterized by a segmented body divided longitudinally into three lobes and transversely into three sections from head to tail. Early Cambrian–end Triassic.

uniformitarianism The principle originating with James Hutton (1726–97), stating that the laws of nature now prevailing have always prevailed and that, accordingly, the results of processes now active resemble the results of like processes of the past.

varanids A group of lizards (monitor lizards). Late Cretaceous–extant.

vertebrates The large grouping which includes all animals with a backbone or spinal column ranging from fish to mammals. Late Cambrian–extant.

bibliography and useful websites

Benton, Michael J *Vertebrate Palaeontology* (Chapman & Hall 1991)

Benton, Michael J *When Life Nearly Died: The Greatest Mass Extinction of All Time* (Thames & Hudson, 2003)

Briggs, Derek EG and Peter R Crowther *Palaeobiology II* (Blackwell Science, 2001)

Clarkson, ENK *Invertebrate Palaeontology and Evolution*, (Blackwell Science, 1998)

Condie, Kent C *Plate Tectonics and Crustal Evolution* (Butterworth-Heinemann, 1998)

Currie, Philip J and Kevin Padian (eds) *Encyclopedia of Dinosaurs* (Academic Press, 1997)

Hancock, Paul L and Brian J Skinner (eds) *The Oxford Companion to The Earth* (Oxford University Press, 2000)

Kingdon, Jonathon *Lowly Origin: Where, When, and Why Our Ancestors First Stood Up*, (Princeton 2003)

Knoll, Andrew H *Life on a Young Planet: the First Three Billion Years of Evolution on Earth* (Princeton, 2003)

Lewin, Roger *Principles of Human Evolution* (Blackwell Science, 1998)

Lewis, CLE and SJ Knell *The Age of the Earth: from 4004 BC to AD 2002* (Geological Society, 2001)

Palmer, Douglas *The Atlas of The Prehistoric World* (Marshall Editions, 2000)

Palmer, Douglas *Fossil Revolution: The Finds that Changed Our View of the Past*, (HarperCollins, 2003)

Press, Frank and Siever, Raymond *Understanding Earth* (WH Freeman & Co 2000)

Van Andel, TH *New Views on an Old Planet: a History of Global Change* (Cambridge University Press, 1994)

Wilson, RCL, Drury SA, and Chapman JL *The Great Ice Age: Climate Change and Life* (Routledge 2000)

The World Wide Web is an excellent source of information about geology. However, as with so much web "data", you have to be careful about the reliability of the information. Generally, those sites set up by major national institutions such as geological surveys, museums, universities and publishers of internationally creditable journals can be relied upon, as can many of the links they provide because they are not trying to sell you anything. I have found the following useful:

University of California Museum of Paleontology, Berkeley www.ucmp.berkeley.edu; **Natural History Museum, London** www.nhm.ac.uk; **The Geological Society, London** www.geolsoc.org.uk; **The Palaeontological Association, UK** www.palass.org, **American Museum of Natural History** www.amnh.org; **British Geological Survey** www.bgs.ac.uk; **Geological Society of America** www.geosociety.org; **National Museum of Natural History, Smithsonian Institution, Washington D.C.** www.nmnh.si.edu/paleo/links.html; **United States Geological Survey;** www.usgs.gov/index.html; *National Geographic* www.nationalgeographic.com; *American Scientist* www.amsci.org/amsci; *Discover* www.discover.com; *Nature* www.nature.com; *Scientific American* www.sciam.com; *Science* www.sciencemag.org

index

acknowledgements

(in page order) 2 Senckenberg, Messel Research Department, Frankfurt am Main; 5 Science Photo Library/Joe Tucciarone; 9 Ministère de la Culture, A Chéné - Centre Camille Jullian - Centre National de Préhistoire, Périgueux; 11 Getty Images/Image Bank/Jeff Hunter; 12 Still Pictures/Jacques Jangoux; 13 NASA; 14 Still Pictures/M & C Denis-Huot; 15 Still Pictures/Fred Bruemmer; 17 Science Photo Library/William/Ervin; 18 top left NASA; 18 top right NASA; 18 bottom Ministère de la Culture, A. Chéné - Centre Camille Jullian - Centre National de Préhistoire, Périgueux; 19 Tony Waltham Geophotos; 21 NASA GSFC Visualization Analysis Laboratory; 23 Tony Waltham Geophotos; 24 Lochman Transparencies/Alex Steffe; 25 AKG-Images/Antiquarium Museum, Pompeii; 26 Tony Waltham Geophotos; 27 Science Photo Library/George Bernard; 29 Douglas Palmer/Dr Phil Lane; 30 Douglas Palmer; 31 Agence France Presse/Ravanelli/Getty Images; 32 Douglas Palmer; 34 Getty Images/Taxi/Harvey Lloyd; 35 M.O. & J Plassard; 36 Geology Institute, Neuchâtel University, Switzerland; 37 Douglas Palmer; 39 Illustrated London News Picture Library; 40 Science Photo Library/Philippe Plailly/Eurelios; 41 Douglas Palmer; 43 The Society of Antiquaries of London; 44 Science Photo Library/Volker Steger/Nordstar - 4 Million Years of Man; 45 Science Photo Library/Tom McHugh/Field museum, Chicago; 46 Tony Waltham Geophotos; 47 Science Photo Library/Dr Juerg Alean; 49 Senckenberg, Messel Research Department, Frankfurt am Main; 50 © The Royal Society, London; 53 © The Natural History Museum, London; 55 Science Photo Library/John Reader; 56 Oxford Scientific Films/Clive Bromhall; 57 Science Photo Library/John Reader; 58 Senckenberg, Messel Research Department, Frankfurt am Main ; 59 Senckenberg, Messel Research Department, Frankfurt am Main; 60 Hessisches Landesmuseum Darmstadt; 62 NASA/Jacques Descloitres, MODIS Land Science Team; 63 top Science Photo Library/Dr Ken Macdonald; 63 bottom Science Photo Library/Dr Ken Macdonald; 64 Science Photo Library/NASA/JPL; 65 NASA/Jacques Descloitres, MODIS Land Rapid Response Team, NASA/GSFC; 67 Science Photo Library/Chris Butler; 69 Science Photo Library/Jim Amos; 71 © The Natural History Museum, London; 73 Science Photo Library/Sheila Terry; 74 Tony Waltham Geophotos; 75 Tony Waltham Geophotos; 77 top Pratt Museum of Natural History at Amherst College; 77 bottom Pratt Museum of Natural History at Amherst College; 78 Science Photo Library/Joe Tucciarone; 79 Museum des Sciences Naturelles de Belgique; 81 © The Natural History Museum, London; 82 © The Natural History Museum, London /M Long; 83 Dr Thomas Martin; 84 Dr David Dilcher; 85 top right Carnegie Museum of Natural History, Pittsburgh/Mark A Klingler; 85 bottom left © The Natural History Museum, London; 87 Professor Michael Hambrey; 89 Science Photo Library/Professor Walter Alvarez; 90 Peter Sheldon; 93 Science Photo Library/Joe Tucciarone; 94 Professor Greg Retallack; 95 Science Photo Library/Dr David Kring; 96 Laszlo Keszthelyi; 97 Science Photo Library; 98 Professor Daniel Schrag; 99 Tony Waltham Geophotos; 101 Douglas Palmer; 105 Science Photo Library/Ted Clutter; 106 Professor Michael Hambrey; 107 Peter Sheldon; 108 Professor Michael Hambrey; 111 The Art Archive/Musée National d'Art Moderne, Paris/Dagli Orti; 112 Science Photo Library/John Durham; 113 Professor Andrew Scott; 114 left Science Photo Library/Worldsat International Inc; 114 right Science Photo Library/Worldsat Productions/NRSC; 119 Reproduced by permission of the British Geological Survey. © NERC. All rights reserved. IPR/41-7C; 120 Dr Jennifer A Clack; 121 Dr Jennifer A Clack; 123 Science Photo Library/Sinclair Stammers; 124 Science Photo Library/Sinclair Stammers; 125 Reproduced with the permission of the Minister of Public Works and Government Services Canada, 2003 and Courtesy of Natural Resources Canada, Geological Survey of Canada; 126 Degan Shu, Northwest University, Xi'an and Simon Conway Morris, University of Cambridge; 129 Professor Michael Hambrey; 130 Reproduced by permission of the British Geological Survey. © NERC. All rights reserved. IPR/41-7C; 131 NASA/GSFC/LaRC/JPL, MISR Team; 132 Reproduced with the permission of the Minister of Public Works and Government Services Canada, 2003 and Courtesy of Natural Resources Canada, Geological Survey of Canada, photo: S McCracken; 133 Lochman Transparencies/Len Stewart; 134 © H J Hofmann; 135 University of California Museum of Palaeontology, Berkeley; 136 David M Rudkin, Royal Ontario Museum; 137 Dr T P Crimes; 138 Professor Michael Hambrey; 141 Tony Waltham Geophotos;142 Corbis/Bettmann; 143 top Science Photo Library/Dr Kari Lounatmaa;143 bottom Tony Waltham Geophotos; 144 Science Photo Library/NASA; 145 NASA; 147 Professor Shuhai Xiao and Professor Andrew H Knoll; 149 Science Photo Library/Space Telescope Institute/NASA; 150 Getty Images/Hulton Archive; 151 Mary Evans Picture Library; 152 Professor Sam Bowring; 155 Corbis/James L Amos; 156 NASA JSC/Carl Allen; 157 top left NASA; 157 top right NASA; 157 bottom left NASA; 157 bottom right NASA; 159 Science Photo Library/R Ermakoff/Eurelios; 161 Science Photo Library/NASA; 162 AKG-Images/Rijksmuseum, Amsterdam; 166 Tony Waltham Geophotos; 167 Tony Waltham Geophotos; 168 Still Pictures/Lo Tsung Hsien/UNEP; 169 Tony Waltham Geophotos